W0017467

Cameroon
Forest Sector Development in a Difficult Political Economy

Evaluation Country Case Study Series

B. Essama-Nssah
James J. Gockowski

2000

The World Bank

Washington, D.C.

www.worldbank.org/html/oed

ISBN 0-8213-4760-8

Library of Congress Cataloging-in-Publication Data

Essama-Nssah, B. (Boniface), 1949-
    Cameroon : forest sector development in a difficult political economy / Boniface Essama Nssah,
James Gockowski.
       p. cm. -- (Operations evaluation study)
    Includes bibliographical references (p.).
    ISBN 0-8213-4760-8
       1. Forest management--Cameroon. 2. Forest policy--Cameroon. I. Gockowski, James Jerome.
    II. Title. III. World Bank operations evaluation study.

SD352.C17 E88 2000
333.75'096711--dc21   00-032516

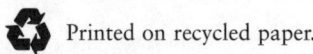 Printed on recycled paper.

Table of Contents

Boxes

Tables

Figures

Foreword

This case study is one of six evaluations of the implementation of the World Bank's 1991 Forest Strategy. This and the other cases (Brazil, China, Costa Rica, India, and Indonesia) complement a review of the entire set of lending and nonlending activities of the World Bank Group (IBRD, IDA, IFC, and MIGA) and the Global Environment Facility (GEF) that are pertinent to the Bank Group's implementation of the forest strategy. Together these constitute inputs into a World Bank Operations Evaluation Department (OED) synthesis report entitled *A Review of the World Bank's 1991 Forest Strategy and Its Implementation*. This forest strategy evaluation was carried out under the overall direction of Uma Lele.

The purpose of each of the six country studies has been to understand the implementation of the 1991 Forest Strategy in Bank operations and to obtain the views of the various stakeholders in the country about the involvement of the Bank. In doing so, the study team has not only examined the Bank's forest program, but also endeavored to place the Bank's activities in the broader context of what the country and other donors have been doing in the forest sector. Therefore, each country study examined the overall development of the country's forest sector. While this naturally includes environmental impacts on forests, such as degradation, biodiversity loss, and deforestation, it also encompasses the economic uses of forests, including the management of forest resources for production, the role of forest development in poverty alleviation, and the impacts of forest research and development.

The evaluation of the Bank's performance in these studies, as always in OED studies, seeks to judge whether the Bank has "done the right things" and "done things right." Here, OED also seeks to judge whether the Bank has lived up to the commitments made in its 1991 Forest Strategy. The case studies do this by examining how the Bank, using the various lending and nonlending instruments at its command, has interacted with the sector's development processes, with other donors, and with the broader government objectives of economic growth, poverty alleviation, and environmental sustainability. Thus, the studies focus on policy in the post-1991 period, but they also recognize that the Bank does not operate in isolation from its historical interactions with a country and its needs. These interactions include the Country Assistance Strategies or their predecessors, economic and sector work, as well as all investments in all sectors and all policy dialogue that is pertinent to the Bank's actions and their outcomes in the forest sector. Together, these activities constitute the Bank's implementation of its forest strategy in a country.

The important questions these country studies address are as follows:

- How have the forces of development effected change in the country's forest sector?
- Did the Bank's 1991 Forest Strategy make a difference to its forest strategy in the country, or was this strategy largely a result of the Bank's historical relationship with the country, the needs articulated by the government, or a combination of both?
- Regardless of how the Bank's forest sector strategy evolved, how consistent was it with the Bank's 1991 Forest Strategy?
- How consistent was the country's own forest policy/strategy with the Bank's 1991 Forest Strategy?
- Was the Bank's overall and forest sector strategy in the country relevant to the country's needs in the forest sector, as identified by the country?
- Were the Bank's overall and forest sector activities effective from the viewpoint of the intentions of its 1991 Forest Strategy?
- Were the Bank's activities efficient?
- Did the Bank's activities achieve policy and institutional development pertinent to forest sector management?

- Are the Bank's impacts likely to be sustainable?
- What impact has the Bank's overall and forest sector strategy for the country had on forest cover and quality, poverty alleviation, and other key issues? What are the prospects for future Bank-country interactions in the forest sector, and for outcomes in the sector?

Gregory Ingram
Director
Operations Evaluation Department
The World Bank

Director-General, Operations Evaluation Department: Mr. Robert Picciotto
Director, Operations Evaluation Department: Mr. Gregory Ingram
Task Manager: Ms. Uma Lele

Acknowledgments

OED's *A Review of the World Bank's 1991 Forest Strategy and its Implementation*, of which this study is a part, was conducted under the direction of Uma Lele. This report is a collaborative effort between B. Essama-Nssah of the Operations Evaluation Department (OED) of the World Bank and James J. Gockowski of the International Institute of Tropical Agriculture in Cameroon (IITA-Cameroon). The collaboration was made easier by two key factors: (1) The congruence of the objectives of the OED review of the 1991 Forest Strategy and the goals of the Alternatives to Slash and Burn (ASB) program of the IITA, and (2) the understanding of Mr. Lukas Brader, Director-General, IITA-Cameroon. This arrangement made it possible for the study to be enriched by the vast field experience IITA has accumulated in Cameroon. OED is grateful to Mr. Brader for making this possible. Dominique Endamana and Guy-Blaise Nkamleu of the IITA Humid Forest Research Station supported the study through diligent efforts in both field and library research. William Hurlbut provided editorial support for this study. The report has benefited from comments on earlier drafts by Messers. Robert Picciotto and Gregory Ingram of OED.

This review has also benefited from field visits by country authors and comments from Bank staff. Mr. Giuseppe Topa of the Africa Region of the World Bank offered excellent cooperation in sharing with the authors all relevant information at his disposal and his insights into the forest sector in Cameroon. He also provided very useful comments on the first draft of the report. That draft benefited also from very insightful comments by Mr. Vicente Ferrer-Andreu of the World Bank Institute (WBI). A sincere thanks to Mr. Eric Chinje of External Affairs, World Bank, and Mr. Laurent Debroux and Ms. Helen Pieume of the Cameroon Resident Mission in helping disseminate the preliminary report to in-country stakeholders for comment.

The team is also grateful for comments that it has received from stakeholders both within and outside the country, including NGOs, government officials, academia, international organizations, and individuals from the private sector. The study was also discussed at OED's Forest Strategy Review Workshop on the Preliminary *A Review of the World Bank's 1991 Forest Strategy and Its Implementation*, held in Washington, D.C., on January 27 and 28, 2000. The team is grateful to Messers. Benjamin Tchoffo of the African Center for Applied Forestry Research and Assene-Nkou, General Secretary of the National Union, and Dominic Walubengo of the Forest Action Network, Kenya, participants at the workshop, for their comments. In addition, comments were received from a web-based consultation on the OED forest strategy review and an ESSD regional consultation on the Forest Policy Implementation Review and Strategy in South Africa, May 3–5. The report also benefited from comments from Messers. William Sunderlin and David Kaimowitz of the Center for International Forestry Research (CIFOR). Thanks are due to Ms. Frances Seymour and Mr. Navroz Dubash of the World Resources Institute (WRI) for their generosity in sharing with OED several iterations of their analysis of the effectiveness of adjustment lending as an instrument for forest policy reforms in Cameroon, Indonesia, and Papua New Guinea. This work has been a valuable input to this report. The team is especially grateful to Mr. Francois E. Ekoko of the United Nations Development Program for his valuable comments. A sincere thanks also to the Rainforest Foundation and the Rainforest Action Network (RAN) for their comments. Some of these comments raised concerns about the environmental consequences of Bank Group–supported activities in the early years of the post-strategy period. However, current safeguard procedures ensure that the environmental and social impacts of Bank Group activities are addressed through a due diligence process.

Since the preparation of this report in October 1999, the Bank staff working in the forest sector argue that some things remain the same— for example, the way the sector is managed, incentives for conservation, and the very limited implementation capacity—but that there is progress in other areas. The government is showing greater interest in

Bank support for forest sector projects. Expectations have been created in the government, the donor community, and the environmental NGOs about the Bank's future support for forest projects in Cameroon. The Bank's forest sector staff argue that this will require a long-term commitment on the part of the Bank, including the use of the entire set of instruments at its command, going beyond the short-term adjustment lending of the past. They believe that there is progress in this regard within the Bank, exemplified by the preparation of the Country Assistance Strategy and the review of the appropriate lending instruments to achieve forest sector objectives. Time will tell if the government commitment to the forest sector has increased and if the Bank will have a more effective role in the Cameroon's forest sector in the future.

The study was published in the Partnerships and Knowledge Group (OEDPK) by the Outreach and Dissemination Unit. The task team includes Elizabeth Campbell-Pagé (task team leader), Caroline McEuen (editor), Kathy Strauss (graphics and layout), Diana Qualls (editorial assistant), and Juicy Qureishi-Huq (administrative assistant).

Acronyms

ASB	Alternatives to Slash and Burn (IITA)
CAS	Country Assistance Strategy
CGIAR	Consultative Group on International Agricultural Research
CIDA	Candian International Development Association
CIFOR	Center for International Forestry Research
CIRAD	Centre de Coopération Internationale en Recherche Agronomique pour le Développment
CPPR	Country Portfolio Performance Review
DRC	Domestic resource cost
EPC	Effective protection coefficient
EPHTA	Ecoregional Programme for the Humid and Sub-humid Tropics of Sub-Saharan Africa
ERC	Economic Recovery Credit
ESSD	Environmentally and Socially Sustainable Development
FAO	Food and Agriculture Organization
FOB	Free on board
GEX	Groupement des Exportateurs
GDP	Gross domestic product
GEF	Global Environment Facility

IBRD	International Bank for Reconstruction and Development (World Bank)
ICR	Implementation Completion Report
IDA	International Development Association
IFC	International Finance Corporation
IITA	International Institute of Tropical Agriculture
IRA	Institut de Recherche Agronomique
IRAD	Institut Rerecherche Agricule pour le Développement
IRZV	Institut de Recherche Zootechnique et Vetrinaire
MDB	Multilateral Development Bank
MED	Minimum exploitable diameter
MIGA	Multilateral Investment Guarantee Agency
MINEF	Ministry of Environment and Forests
NDVI	Normalized difference vegetation index
NEMP	National Environmental Management Plan
NGO	Nongovernmental organization
NMB	National Marketing Board
NPV	Net present value
OED	Operations Evaluation Department
ONCC	Office National du Café et Cacao
ONCPB	Office National de Commercialisation des Produits de Base
OP	Operational Policy
PAD	Project Appraisal Document
PAM	Policy analysis matrix
PAR	Performance Audit Report
PNVA	Programme National de Vulgarisation Agricole
QAG	Quality Assurance Group
SAC	Structural Adjustment Credit
SAL	Structural Adjustment Loan
SAR	Staff Appraisal Report
SDA	Subregional delegate of agriculture
SDF	Social Democratic Front

SODECAO	Société de Développement du Cacao
UNDP	United Nations Development Program
UPC	Union des Populations du Cameroun
WRI	World Resources Institute
WWF	World Wide Fund for Nature

Summary

This country case study, part of the Operations Evaluation Department (OED) *A Review of the 1991 World Bank Forest Strategy and Its Implementation*, evaluates Bank operations in Cameroon for their consistency with the strategy. The strategic aspects of those operations are judged here on their relevance, effectiveness, efficiency, institutional development, and sustainability.

The forests and biodiversity of Cameroon constitute a significant portion of the Congo Basin. The Congo Basin accounts for about 80 percent of the remaining moist forests in Africa and 20 percent of the world's tropical moist forest—second in size only to the Amazon. The Bank's 1991 Forest Strategy listed all the countries of the Congo Basin (except Equatorial Guinea) among the 20 countries with threatened tropical moist forests, implying that these countries deserved special attention in its programs. Since that time, the timber sector in Cameroon has significantly increased its logging activities and made Cameroon the leading exporter of tropical timber in Africa. Although the Bank has undertaken little direct forest investment over the past decade, forest sector issues have been part of the Bank-financed Structural Adjustment Program since 1989.

Forests and Forest Policy in Cameroon

As in many other developing countries, there are few reliable sources of statistical information about forest loss in Cameroon, but estimates of annual deforestation range from 0.4 to 1.0 percent. The causes of deforestation are similarly uncertain. While smallholder slash-and-burn agriculture and fuelwood demand are widely believed to be responsible for about 90 percent of the deforestation, these factors are often secondary effects of tropical timber harvesting that degrades forest cover and contributes to associated declines in biodiversity. The four dynamic mechanisms of forest resource degradation and deforestation described in the 1991 Bank strategy—shortened fallow cycles, direct conversion, logging, and fuelwood demand—are all applicable to the Cameroon case. Whatever the primary cause, however, the current levels of deforestation and biodiversity degradation are not efficient, equitable, or sustainable.

Cameroon's forest sector experience is strongly linked to events in agriculture and the overall political economy. The low productivity of the agricultural system, combined with increased food demand, has made the expansion of the cultivated area a leading cause of deforestation. Bank-supported, policy-led attempts by the government of Cameroon to intensify agricultural production were largely unsuccessful, primarily because of poor institutional development at both the grassroots and the central levels and an inappropriate macroeconomic framework (particularly prior to the 1994 devaluation of the CFA franc).

The fundamental objective of the forest policy reform in Cameroon was to establish a transparent, equitable, and sustainable management system for forest resources. The outcome of the reform process was limited, for four reasons. First, the government of Cameroon lacked genuine commitment and the capacity to carry out the reform. Second, key actors in the reform process (particularly foreign logging companies and the parliament) chose to oppose it. Third, partners such as the World Bank failed to devise an implementation strategy compatible with the underlying dynamics of political and socioeconomic changes in Cameroon. Finally, while Cameroon's forest policy is well codified in documents, it is poorly implemented.

Although the reforms have led to increased tax revenues and increased the share of GDP attributable to the forest sector, the structural underpinnings of the sector have been little affected. Government agencies in the sector continue to be weak. The international logging companies that dominate the sector continue to have a free hand in the development and use of the forest resources of Cameroon. Local communities were left out of the reform process, despite the declared objective to include them in forest resource management.

The Bank's Performance

During the 1980–99 period, direct World Bank involvement in the forest sector of Cameroon consisted of a plantation project in 1982, ongoing policy-based lending that started in 1989, and a Global Environment Facility (GEF) biodiversity project approved in 1995. Indirect lending has consisted primarily of a declining number of agriculture projects directed at the primary export crops of Cameroon, including cocoa and other tree crop plantations. The forest policy advice of the Bank to the government of Cameroon was formulated with the 1987 Tropical Forestry Action Plan, the 1989 review of the agriculture sector, and Country Assistance Strategies in 1994 and 1996.

Among the things that the Bank helped achieve in Cameroon are the following: (1) forest sector issues are now at the center of the policy debate; (2) a multisectoral approach has been adopted to the extent that forest policy reform has been a key feature of the Country Assistance Strategy process and the Structural Adjustment Program; (3) Bank interventions have focused on areas recommended by the 1991 Forest Strategy—policy and institutional reforms; and (4) cocoa and coffee marketing have been liberalized. With respect to international cooperation, the Bank and the rest of the donor community played a significant role in the outcome of the Yaoundé Declaration, which commits the governments of the Congo Basin to sustainable forest management.

However, the Bank did not attempt all the right and relevant things prescribed by the 1991 strategy. In particular, the Bank abandoned rural development in Cameroon and thereby missed an opportunity to promote the participation of rural people in resource expansion and intensification. In the 1992–99 period, because of lack of government commitment and weak capacity, the Bank shifted its lending approach in Cameroon from agriculture to multisector[1] activities and to public sector management based on adjustment lending.

Over time, poverty increased as the economy declined. The government of Cameroon bears the primary responsibility for this outcome. It failed to consistently articulate a vision of socioeconomic development compatible with poverty reduction and environmental sustainability, and going beyond the diverse private interests in society. When it proclaimed a vision that would satisfy the World Bank and other partners, it failed to back up such declarations with credible actions that would rally all stakeholders around the policy objective, specify and enforce jurisdictional boundaries among government agencies, and marshal the resources (including building up the necessary capacity) to do the job.

The Bank made several strategic mistakes in Cameroon. First, it relied too heavily on the executive branch of the government to deliver on the promised reforms. Second, in both agriculture and forests, the Bank neglected the creation and dissemination of knowledge and information that was crucial for policymaking and implementation. Reliable information, when available to all the key players, reduces transaction costs associated with policymaking. Third, the Bank rightly recognized institutional weaknesses in Cameroon, but preferred to rely heavily on technical assistance to deal with the issue. Failure to develop local institutions undermined the sustainability of any achievements. Finally, the Bank had good intentions in trying to promote the interest of local communities, but it did little to gather their views and to design mechanisms that would ensure that those views were taken into consideration. The Bank should have made a truly participatory approach a cornerstone of the policy-based lending program.

Overall, the interventions of the Bank inside and outside the forest sector in Cameroon were relevant to its strategic objectives, but they were neither efficacious nor efficient. Because of weak institutional development, the achievements are unlikely to be sustained. The Bank should focus its future reform efforts in Cameroon on the collection and dissemination of relevant and reliable information, working with a larger set of stakeholders, and using more Cameroonian expertise to gain local perspective and build capacity. The success of such an approach hinges on government commitment and the cooperation of other donor countries, including those with timber interests in Cameroon.

1

An Overview of the Forest Sector in Cameroon in the Context of the 1991 World Bank Strategy

The Forest Sector: A World Bank Policy Paper (1991) articulated a major revision in Bank forest strategy that addressed concerns about the environmental effects of its projects and policies on moist tropical forests in developing countries (see box 1.1). This case study examines the impact on the management of Cameroon forest and woody biomass resources of World Bank activities and policy recommendations before and after the 1991 revision of Bank strategy (see box 1.2 for a description of the OED review of which this study is part).

There are two major environmental concerns facing policymakers and stakeholders concerned with the moist forests of Cameroon: deforestation and forest degradation. Smallholder agriculture is allegedly responsible for 85–95 percent of total deforestation in Cameroon and is one of two major foci of the study; the other is the forest degradation and sustainability of tropical timber harvesting. The Bank, beginning with its first mission in 1964 until the late 1980s, viewed Cameroon's forests as an unused resource. This view, although more moderate today, is still pervasive in Cameroon and underlies the rapid expansion of the forest sector in the last 15 years. Given that some deforestation and forest degradation are bound to take place in the course of economic development, the question then is: *When do these phenomena become a policy issue?* In principle, these processes become problems to be dealt with

Box 1.1. Bank Forest Strategy: The 1991 Forest Paper and the 1993 Operational Policy Directive

The 99-page World Bank publication *The Forest Sector: A World Bank Policy Paper* was published in September 1991. This paper (henceforth referred to as the 1991 forest paper) represented the initial comprehensive statement of a new direction for the Bank's forest strategy. A two-page Operational Policy directive (OP 4.36, produced in 1993) reflected the policy content of the paper, and a Good Practices summary (GP 4.36) provided operational direction to Bank staff. The 1991 forest paper, the OP, and the GP are together the subject of OED's evaluation.

In today's Bank terminology, the 1991 forest paper sets out a Bank strategy and the OP defines the policy. The 1991 forest paper gave guidance on policy directions, programmatic emphases, and good practice, and it specified principles and conditions for Bank involvement in the forest sectors of its client countries. It was the first instance of significant outside stakeholder participation in the formulation of a Bank sector strategy, and it is this document which the public considers the embodiment of the new direction for the Bank's forest strategy. Both the Bank's Board and civil society were referring to this document, as well as OP 4.36, when they asked OED for an independent evaluation of the Bank's forest policy. Although the Foreword for the 1991 forest paper was signed by then Bank President Barber Conable, the Board was not asked to, nor did it, comprehensively approve the 1991 forest paper. However, it did discuss the paper and endorse specific aspects of it.

The Board-endorsed principles contained in the 1991 forest paper included the ban on financing commercial logging in primary topical forests; incorporation of forest sector issues into the general policy dialogue and country assistance strategy; and promotion of international cooperation, policy and institutional reform, resource expansion, and forest preservation. The endorsed principles also included the statement that "in tropical moist forests the Bank will adopt, and will encourage governments to adopt, a precautionary [sic] policy toward utilization ... Specifically, the Bank Group will not under any circumstance finance commercial logging in primary tropical moist forests. Financing of infrastructural projects... that may lead to loss of tropical moist forests will be subject to rigorous environmental assessment as mandated by the Bank for projects that raise diverse and significant environmental and resettlement issues. A careful assessment of the social issues involved will also be required" (p. 19). The Board also approved a specific section on conditions for Bank involvement.

Both the 1991 forest paper and the OP emphasize that the Bank will not finance commercial logging in primary tropical moist forests, and in addition, the 1993 OP adds that the Bank "does not... finance the purchase of logging equipment for use in primary tropical moist forests" (para. 1a). The OP also states that "in areas where retaining the natural forest cover and the associated soil, water, biodiversity, and carbon sequestration values is the object, the Bank may finance controlled sustained-yield forest management" (para. 1f). The 1991 paper, however, had stressed a lack of agreement on what constitutes sustainable forest management and offered three different definitions of it. However, all definitions of sustainable forest management typically include management of forests for *multiple uses* as distinct from timber production alone, to which logging normally refers. Although this provision in the OP to finance forest management under controlled sustained-yield conditions allows forest management under specific conditions (and the drafters of the OP thought this introduced some flexibility for the Bank), a survey indicates that the staff have not considered the OP to be flexible on this point. The Bank will need a clearer policy if its future lending and nonlending activities are to address issues of improved forest management relative to current logging practices in many countries, which this report argues often tend to be environmentally destructive and socially inequitable. What constitutes "sustainable" forest management will, in all likelihood, remain unresolved and specific to each location.

Based on the larger policy statement, the OP also states that "the Bank distinguishes investment projects that are exclusively environmentally protective... or supportive of small farmers... from all other forestry operations." It goes on to say that projects in the latter category "may be pursued only where broad sectoral reforms are in hand, or where remaining forest cover in the client country is so limited that preserving it in its entirety is the agreed course of action" (para. 1c). The main report for this study finds that the Bank could more usefully and proactively work with stakeholders sympathetic to reforms in borrowing countries in ensuring that reforms are in hand, rather than wait for them to occur before getting engaged in the forest sector.

Box 1.2. The Operations Evaluation Department Review of the 1991 Forest Strategy and Its Implementation

OED's review of the Bank's 1991 Forest Strategy[1] has been undertaken to assess Bank experience in the forest sector—particularly since 1991—to gauge its policy intentions, implementation, and impacts. The review also examines whether the Bank's strategy remains relevant and can embrace a strategy attuned to the current realities of the forest sector. In addition to briefing the Bank's Board of Executive Directors, the review will be used as an input to an ongoing Bank-wide review of its forest sector activities being lead by the Bank's Environmentally and Socially Sustainable Development Network (ESSD).

Cameroon was chosen as a case study for several compelling reasons. Since the onset of the 1990s, the timber sector in Cameroon has significantly increased its logging activities, to the point where the country is now recognized as the leading exporter of tropical timber in Africa. Another important reason for studying Cameroon is the importance of its biological diversity, particularly in its moist forests, which are considered to be some of the world's most diverse tropical rainforests (see Annex B). Cameroon has also been participating in a World Bank structural adjustment program since 1989. Finally, the relatively intact eastern frontier forests of Cameroon form the western portion of the world's second-largest tropical forest biome, the Congo Basin. Thus, lessons from the Cameroonian experience can prove useful in developing strategies for conserving the resources of this last great rainforest for the benefit of present and future generations.

All of the case studies in this review consist of two parts—the first focusing on the extent and causes of changes in the forest sector, and the second on how the entire set of Bank instruments has interacted with the processes of the changing forest cover, and with what impact.

To the extent possible, the performance of the Bank has been assessed based on outcomes and impacts. Six classes of outcome are considered:
- Improvement in country policies and strategies with direct and indirect impacts on forests
- Institutional development including improvement of the legal framework, a redistribution of roles between the public and private sectors, and participatory approaches to decisionmaking
- Improvements in technologies
- Capacity building and human capital formation
- Improvement in the incentive structure
- Improved information, monitoring, and evaluation systems.

1. The strategy is summarized in Annex A.

when they are taking place at rates that are considered socially, environmentally, or economically undesirable. Under these circumstances, the key underpinning issue usually stems from a disagreement over the most appropriate uses of the forestlands (Gregersen et al. 1994: 138).[2]

In examining the moist forest–agriculture interface, the study draws heavily on the findings of the Alternatives to Slash and Burn Program in southern Cameroon and the work of the International Institute of Tropical Agriculture (IITA) and the Center for International Forestry Research (CIFOR) in Cameroon,[3] where exhaustive characterization of forest and agro-ecosystems has been ongoing for the past four years.[4] Stakeholder interviews were also conducted with ministry officials, Bank mission personnel, agricultural extension workers, agricultural and forestry researchers, and community members in 20 villages across southern Cameroon.

The Nature of the Challenge

One of the major challenges facing Cameroonian society is resolving the overexploitation that results when private costs of forest resource use (mainly labor, capital, and transaction costs) do not include all the social costs. The excluded costs include soil erosion, loss of bio- and cultural diversity, greenhouse gas emissions, and watershed degradation—all within the context of the potential irreversibility of tropical deforestation.[5] When external production diseconomies are not reflected in the prices for forest products, there will be little incentive for forest regeneration and sustainable forest management. Redressing the divergence between private and social costs is recognized in the Bank's 1991 strategy. But apart from the GEF-funded biodiversity project, the Bank portfolio in Cameroon did not address these issues.

Deforestation and Degradation of Moist Tropical Forests in Cameroon

Moist tropical forests provide a range of products and services, including timber and non-timber forest products, the utilization of forest biomass as a fertility input (when converted to ash through slash-and-burn techniques), the conservation of important biodiversity, the protection of soil resources and watersheds, the prevention of desertification, and the regulation of local and global climatic patterns through carbon sequestration.

One of the chief conundrums of the agriculture-forest interface is the inherent conflict between the fertility function of forest biomass in slash-and-burn agriculture and the other forest functions that are compromised when forestland is cleared for agriculture. Resolving this and other problems in managing the resource requires accurate measuring, monitoring, and social valuation of forest functions.

However, a major problem in Cameroon and most of tropical Africa is the lack of reliable and consistent statistics on the extent of forest cover and condition and its changes over time. From the various existing estimates of deforestation in Cameroon, politicians, environmentalists, and other concerned stakeholders are able to choose the estimate best suited to supporting their position. Estimates of the extent of closed-canopy moist tropical forests range from 155,000 km^{-2} (33 percent of national territory) to 206,000 km^{-2} (44 percent of national territory). At the same time, estimates of the rate of deforestation in the 1980s and 1990s range from 800 km^2 annually to 1,500 km^2, which translates into annual rates ranging from 0.4 percent to 1.0 percent (see Annex C). In an interesting approach, Gaston et al. (1998) combine data on carbon pool estimates for forest types with Food and Agriculture Organization (FAO) forest cover maps from 1980 and 1990 and population density data to estimate a 1.7 percent annual decline in total carbon pools due to deforestation and degradation.

Given the global importance of moist tropical forests and the considerable variation in estimates of vegetation cover and change, it is imperative that a reliable and indisputable international monitoring mechanism be developed.[6] The rapid evolution in remote sensing technologies offers the best potential for quantifying global, national, and regional patterns of biomass and forest change (Prince and Goward 1995; Gaston et al. 1998). Reliable and replicable estimates from such techniques would be of great use to policymakers and other stakeholders. For instance, signatories of the UN Framework Convention on Climate Change (e.g., Cameroon) must inventory their sources and sinks of greenhouse gases, including carbon (as CO_2) from changes in land use. For most tropical countries the availability of data for making carbon flux estimates from changes in land use are limited, making the global convention rather meaningless. There is a clear need for a concerted international effort to improve this critical global monitoring.

In designing a strategy, it is important to consider that forest degradation is more difficult to monitor than deforestation, as it is generally unobservable using satellite imagery. The more important degradation issues of forest/woody biomass resources include: (1) the long-run sustainability of logging operations; and (2) the secondary impacts of logging, agriculture, and population growth on biodiversity resources. Successfully addressing these issues will require a strong political commitment to the provisions for sustainable forest management in the new forest strategy, often in the face of opposition from a strong foreign-dominated logging lobby that is viewed by many as not having the long-run sustainable development of the forest sector as a main objective.

Causes of Deforestation and Degradation

In Cameroon, agriculture is allegedly responsible for the lion's share of deforestation. The 1991 Forest Strategy notes that among all countries with tropical moist forests, approximately 60 percent is lost to agricultural settlement, with the balance split roughly between logging and other uses. In Cameroon, agriculture's share is commonly cited as 90 percent. As noted above, a distinction should be made between deforestation and degradation, especially as the actual rate of deforestation in Cameroon may not be extreme. However, the increase in degradation associated with the rapid growth of the logging industry in recent years is of growing concern.

The four dynamic mechanisms of forest resource degradation and deforestation described in the 1991 Bank Strategy can all be found to varying extents in Cameroon: shortened fallow cycles, direct conversion, logging, and fuelwood demand.

Shortened Fallow Cycles

Shortened fallow cycles are considered the most common source of deforestation in Cameroon and are attributed to smallholder agriculture. As population pressures increase and fallow periods become shorter, fallow composition changes from secondary forest pioneer species such as *Terminalia* spp. and *Musanga* spp. to shrubs and grasses. Ultimately, the succession process of the natural forest may be endangered. Increased annual food cropping over time threatens the integrity and viability of the forest ecosystem and, with sufficient population pressure, transforms the landscape into a cultivation/forest mosaic, as now found in most of the former forested areas of Côte d'Ivoire and Ghana. However, there are sustainable pathways for rural development in the humid forest zones that can minimize the damage and, in some cases, even improve the environmental services of the cultivation/forest mosaic ecosystem. The efforts of the Bank to intensify smallholder agriculture in the humid forest zone and its periphery was more emphasized in the 1970s and 1980s than in the 1990s and involved support for extension, agricultural research, integrated rural development, and the cocoa sector.

Direct Conversion

Whereas smallholder agriculture results in a gradual transformation of the landscape as fallows shorten over generations, large-scale plantations are created following the direct conversion of forestlands using mechanical or manual techniques of site preparation to remove existing forest vegetation. In the 1960s, 1970s, and continuing into the early 1980s, Bank-funded projects in the southwestern portion of country focused on the development of tree crop plantations. From 1967 to 1985, nine industrial tree crop projects were approved. Since 1985 there have been no examples of such lending. All of these parastatal enterprises either are now in the process of being privatized or are already sold. Unfortunately these conversions (totaling over 100,000 ha) occurred in the Atlantic coastal forests around Mount Cameroon and south of Douala, which are now recognized as among the world's most biologically diverse tropical rainforests (see Annex B). These efforts can also be criticized on equity grounds, especially when compared to outcomes under smallholder strategies for export crop development.

Logging

The logging process is estimated to open 5 to 10 percent of the forest canopy. This low figure is due to the relatively low harvest rates per hectare (approximately 7 m^3). But with somewhere around 350,000 ha of concessions logged annually, the area deforested would range from 175 to 350 km^{-2}, depending on the actual amount of damage. This would seem to indicate that logging is responsible for a larger overall share of deforestation than is usually reported (these amounts would correspond to between 12 percent and 41 percent, depending on the estimate of total forest change reported above). Again, the statistics are not clear. Much of this deforestation is due to the creation of logging roads, and clearance for yarding and other operations. These intrusions and the influx of laborers and other people also lead to significant collateral degradation to forest resources. Some of this damage involves neighboring communities opportunistically using logging roads to gain access to new lands.

In an interview with a logging company owner operating a concession near the Gabon border, he indicated that extensive areas along the logging tracks were converted by local farmers to plantain and cocoyam fields with the intention of exporting their production to Gabon. Other logging owners interviewed downplayed this impact, noting that because population densities were low and the logging tracks not maintained by the state after logging was finished, there was little linkage with agriculture. These mixed findings were confirmed by rapid rural assessments in 10 southern Cameroonian villages lying adjacent to active logging concessions, which uncovered only a few fields of plantains along newly opened logging roads. Permanent settlements along these logging tracks by either members of the village or by outsiders had

Box 1.3. The Bushmeat Trade in Cameroon

One of the most important non-timber forest product activities within the moist forest zone of Cameroon is the poaching of bushmeat by market hunters. Operating in a close symbiotic relationship with logging concessions, they place indiscriminate snares along newly created logging tracks in order to supply urban markets and the logging camps with both smoked and fresh game. Among the animals entering this trade are such charismatic fauna as elephants, chimpanzees, and gorillas, although the bulk of the trade is based on the duiker, a small forest antelope. The disturbance of important seed spreading fructivores, such as the forest elephant and the black-casqued hornbill (*Ceretogymna atrata*), can have serious repercussions on the forest ecology by negative impacts on these crucial seed dispersal mechanisms.

A household consumption survey conducted in Yaoundé found that the estimated annual consumption of bushmeat exceeded $4 million. No differences in expenditures across income classes were noted, although the poor consumed mainly the cheaper smoked product. For the poor, bushmeat purchases were the second most frequent meat purchase after beef.

not occurred in any of the villages visited. Cameroon has not encouraged the policy pursued by Côte d'Ivoire, where migrants were actively encouraged to convert the logged-over forest to cocoa plantations, and as a result deforestation due to a symbiotic effect between logging and agriculture has not been a major factor (Ruf 1995). Strong traditional customary property rights have also been a deterrent to migrants from outside the forest zone (Diaw 1997). In Cameroon, most collateral degradation is associated with the illegal bushmeat trade (see box 1.3).

Of more concern is the current mode of forest exploitation and its sustainability. First, it should be noted that very little is known about the sustainable management of tropical forests. Preliminary research findings by IRAD (Institut Derecherche Agricole pour le Développement), CIRAD, and Tropenbos in Cameroon on sustainable forest management indicate that the costs of sustainable management are likely to be high and knowledge-intensive. The Bank, using SAL (Structural Adjustment Loan) conditionalities, was instrumental in pushing for the inclusion of sustainable forest management in the 1994 Forestry Law. Given limited resources for effective enforcement and weak financial incentives for sustainable management practices, it is unlikely that timber companies will adopt these practices. A complaint heard from both parliamentarians and Ministry of Environment and Forests (MINEF) officials interviewed was that the *Bank influenced policy, but then made no provisions for implementation or enforcement of those provisions.* Thus, the curtailment of Bank support for the harvesting of tropical timber in primary forest prevents it from playing an active role in the implementation of sustainable practices.

Fuelwood Demand

The demand for fuelwood increased significantly as incomes plunged with the economic crisis and following the doubling in price of traded fuels after the 1994 devaluation. Unfortunately, neither the Bank nor the government made provisions to diminish this added incentive to deforest. A survey of fuelwood consumption conducted by IITA and CIFOR among urban households in the Center and South Provinces of Cameroon found 48 percent and 71 percent of households, respectively, citing it as their principal cooking fuel, compared to 30 percent and 55 percent in 1987 (table 1.1). The impact of fuelwood demand on forest resources is limited, as subsequent studies by CIFOR show that most fuelwood used is a by-product of clearing for agriculture. The fuelwood and charcoal market is the largest forest products market in terms of physical volume of timber felled.[6] Estimated annual per capita consumption for Cameroon is approximately 1 m^3, with total national consumption estimated at approximately 13 million m^3 (Millington and Pye 1994). With market prices in Yaoundé currently averaging 8–15 CFA franc/kg, the estimated value of fuelwood consumed in Yaoundé ranges somewhere between 4.5 billion and 8.4 billion CFA francs annually (US$7.5 million to US$14 million), while the overall market is estimated at between US$45 million and US$65 million.[8] These markets are informal and not subject to government taxation or regulation.

Incentives

The Bank's 1991 Forest Strategy discusses four incentives to deforest that drive the agents and processes discussed above: (1) Increasing population pressures on the natural resource base; (2) Deterioration of income opportunities in settled agricultural regions, leading to increased migration and encroachment on forested land; (3) Increased access to the forest frontier because of infrastructural development; and (4) Introduction of subsidies for non-forest land uses and logging to encourage frontier settlement.

Table 1.1. Percentage of Population Using Fuelwood and Charcoal as Principal Cooking Fuel in 1987

	Total	Rural	Urban
Cameroon	83	96	60
Center Province	60	92	30
South Province	84	94	55

Source: *Demo 87*, volume II, Direction Nationale du Deuxième Recensement Général de la Population et de l'Habitat.

Increasing Population Pressures on the Natural Resource Base

This incentive has been an important factor in Cameroon, which has seen its population grow rapidly at annual rates approaching 3 percent since independence in 1960. However, it is important to examine in detail where this growth has occurred and with what result for the natural resource base, particularly forest resources. Data from the population census of 1976 and 1987 indicate a rapid growth in the urban sector and significant rural-urban migration (table 1.2). In the forest zone, the urban population already exceeded rural population by 1987, at which time it was growing almost 5 times faster. Only the Southwest Province among the forest provinces exceeded the national rural growth rate, reflecting the high influx of migrants from the adjacent and densely populated West and Northwest Provinces attracted by the availability of fertile land. Overall the rural population densities in the forest provinces are quite low (41 percent lower than the national average). The reasons for these generally low densities are numerous. Among the factors are a

Table 1.2. Population Across Agro-Ecological Zones in Cameroon, 1987

	Rural popula-tion	% of total rural	Rural density (pers./ km^2)	Urban popula-tion	% of total urban	Annual rural growth rate (1976–87)	Annual urban growth rate ('76–87)
Savanna provinces	2,170,770	37	13.2	694,174	20	1.8%	7.4%
Adamoua	281,352	5	4.1	159,378	5	1.0%	5.8%
Far North	1,345,756	23	39.3	332,058	10	1.2%	8.6%
North	543,662	9	8.0	202,738	6	3.9%	7.0%
Highland provinces	1,675,434	29	53.7	588,144	17	0.8%	4.6%
Northwest	844,171	14	48.8	234,805	7	0.7%	4.9%
West	831,263	14	59.8	353,339	10	0.9%	4.4%
Forest provinces	1,993,723	34	7.4	2,190,184	63	0.9%	4.4%
South	246,841	4	5.2	95,776	3	0.1%	5.6%
Southwest	491,079	8	19.7	236,047	7	2.0%	2.1%
Center	707,269	12	10.3	765,156	22	0.3%	5.6%
East	308,718	5	2.8	125,943	4	1.1%	5.3%
Littoral	239,816	4	11.9	967,262	28	0.9%	4.0%
Total	5,839,927	100	12.5	3,472,502	100	1.2%	5.0%

Source: Compiled from 1976 and 1987 population census results.

higher incidence of disease (including trypanosomias) and lack of rural communications, in conjunction with economic factors.

Before 1987, there was no rural population growth in the Center or South Provinces. Rural populations in some of the administrative divisions of the South and Center Provinces actually declined from 1976 to 1987. So, in those provinces, there was no direct population pressure on forest resources. But this changed drastically after 1987. Population pressure increased in the Center and the South Provinces, both because of lower rural-to-urban migration and return migration. This was the result of both the economic crisis that started in 1986 and the policy response to it, which hit the urban areas harder that the rural ones.

Among the economic factors posited as influencing rural-urban migration in Cameroon are agricultural terms of trade, urban wages, and the probability of finding urban employment. As seen in figure 1.1, the agricultural terms of trade for the humid forest zone in Cameroon declined progressively from the onset of oil production in the Bight of Biafra in the late 1970s. This decline, in conjunction with the burgeoning civil service (funded by oil royalties) and increasing non-traded good demands in the urban sector, led to a significant exodus of rural population to urban centers. The decline in the terms of trade continued with the economic crisis and did not improve until the devaluation of the CFA

Figure 1.1. Indices of Agricultural Terms of Trade

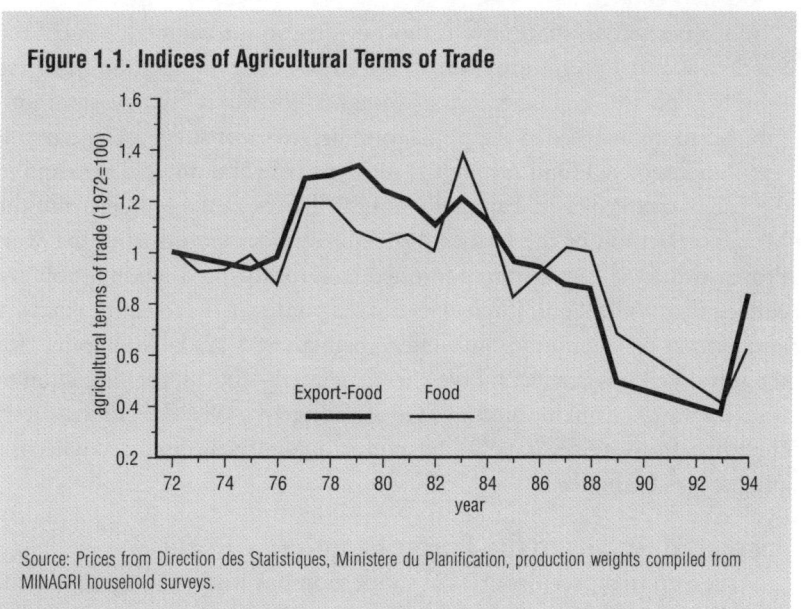

Source: Prices from Direction des Statistiques, Ministere du Planification, production weights compiled from MINAGRI household surveys.

franc in 1994. Improvement in agricultural terms of trade has continued, with higher world prices of coffee and cocoa in 1997 and 1998 and, in combination with reduced probabilities of urban employment and lower urban wages, a slowdown in the rate of rural-urban migration has been noted (Gockowski et al. 1999; Sunderlin, Pokam, and Wadja 1998).

In general, the demographic shift in rural-urban migration is a result of the changing incentive structure facing farmers in the humid forest zone of Cameroon under the newly liberalized market context of structural adjustment. The subsequent impact on forest resources will depend, in the long run, on the types of production systems that are expanded and the intensification process. But in the short run an increased number of households must rely on the same basic production systems and technology that once served a smaller population.

Income Opportunities in Settled Agricultural Regions Have Deteriorated, Leading to Increased Migration and Encroachment on Forested Land

The hypothesis behind declining income opportunities in rural areas is linked to the presumption that rural poor tend to deforest due to necessity.

The private discount rate may be too high. Poor people who through encroachment are responsible for a significant share of the annual loss of forests have higher discount rates than society as a whole because of the urgency of their current needs. (World Bank 1991a: 35).

In Cameroon, this incentive is most evident in the forested areas of the Southwest and Littoral Province where rural-to-rural migration from the densely populated highlands of the West and Northwest Provinces has contributed to an increase in the population pressures of smallholder agriculture. Studies have found the highest incidence of rural poverty here and in the Far North Province. Especially notable has been the settlement of the Moungo Division of the Littoral Province by populations from the West Province and the conversion of forested land for the production of robusta coffee. The World Bank implemented two phases of a project to intensify agricultural production in the western highlands, a laudable objective for the secondary environmental benefits generated—that is, the deflection of rural migration from forested areas to the adjacent lowlands. Unfortunately these benefits were never measured or even qualitatively discussed when the project was evaluated in the 1980s.

Increased Access to the Forest Frontier

Access to the forest frontier in Cameroon has increased greatly in the past 20 years with several road and bridge constructions. In particular, the installation of road infrastructure between the major port of Douala and the East Province has significantly improved access to the rich timber resources of the East Province and was instrumental in its becoming the leading province for timber harvesting. The Fifth Highway Project Loan of the World Bank constructed the Edéa-Yaoundé link, which previously was served by a poorly maintained laterite road that was often impassable during the rainy season. The second major infrastructure project to the east was the continuation from Yaoundé of the paved, two-lane trunk highway to the town of Ayos on the border of the East Province funded by the European Community. The opening up and greater exploitation of the frontier forests of the East Province has had a serious impact on populations of large primates and forest elephants and is threatening the livelihoods of the indigenous Baka forest dwellers. For this the European Union has been heavily criticized by international environmental nongovernmental organizations (NGOs) and is in the process of revising its policy vis-à-vis road infrastructure. However, most Cameroonians applaud the development of this infrastructure and disdain the lobbying of the environmental NGOs to block further donor-funded infrastructure projects. These improvements in road infrastructure have also increased market access for agricultural goods and have led to greater market-led deforestation for agricultural purposes along the trunk roads. This pattern is clearly visible in remote sensing images of southern Cameroon (Thenkabail 1999).

Subsidies for Non-Forest Land Uses and Logging Have Been Introduced to Encourage Frontier Settlement in Some Countries

Direct subsidies for alternative land uses in frontier settlements, which have been widely used in Latin America, have not played a significant role in determining the spatial distribution of populations along the forest frontier in Cameroon. In the 1970s and 1980s, some new planting subsidies for the creation of coffee plantations were paid to both arabica and robusta coffee producers in the west, but these producers were typically not living along the forest frontier margins. Provisions were also made for such new planting subsidies in a World Bank cocoa rehabilitation project for the Center and South Provinces in the early 1990s. These subsidies, however, never saw the light of day.

The Importance of Forests in the Economy of Cameroon

The opening line of the 1995 Forest Policy document of Cameroon notes that "the forests of Cameroon represent one of the country's greatest riches." Overall, ONADEF estimates that the national stock of commercial timber is 310 million m³, which at current FOB prices represents a standing value somewhere in the neighborhood of 25,000 billion CFA francs (approximately US$70 billion). The average productivity of the typical Cameroon production forest is about 260 m³/ha of standing timber, of which 32 m³ are composed of the 75 or so commercially exploited species with 21 m³ exceeding the minimum exploitable diameter[9] (CIRAD-Foret 1997). According to the national zoning plan, 6,025,000 ha are planned as production forests and it is estimated that an average of 415,000 ha of concessions were legally opened for logging from 1994 to 1996 (Eba'a Atyi 1998). The estimated average felling is 7 m³/ha, or less than one tree per ha of production forest—a very low rate compared to other countries producing tropical timber.

Transforming this vast potential wealth into sustainable development presents a number of difficult issues. In contrast to the wealth generated in smallholder agriculture, which contributes directly to more than 2 million livelihoods, wealth from tropical timber is concentrated in the hands of a much smaller group of economic agents. The structure of the industry, ownership patterns, industry investment, employment, and linkages with the rest of the economy are, therefore, key elements that will determine whether sustained and equitable development results. Maintaining these revenue streams requires sustainable management practices, which encounter a host of difficult technical issues.

Structure of the Logging Industry in Cameroon

Historically, logging in Cameroon has been an enclave sector dominated by foreign ownership with high capital requirements and limited forward and backward linkages to the rest of the economy. Nearly all capital input employed in the sector is imported, from chainsaws to heavy equipment. Formal employment in the sector, which includes primary processing firms (sawmills, veneer and plywood factories), is estimated at 33,000 workers, 92 percent of whom are low-paid laborers (Eba'a Atyi 1998). Although the export of raw logs generates the largest portion of sector revenues, it only employs 30 percent of the forest work force, whereas 70 percent are employed by the transformation and processing industries (*Afrique Agriculture* 1990).

Foreign ownership tends to dominate the processing and export sectors and is relatively concentrated. The six largest exporters, all European, account for more than 50 percent of the total export volume and, overall, foreign firms captured two-thirds of total foreign exchange earnings in 1996 (Eba'a Atyi 1998). To address foreign ownership concerns in the forest sector, government policy has sought to increase the participation of national actors. The credit and technical assistance provided by the Canadian International Development Agency (CIDA) to Cameroonian forestry enterprises was a catalyst in the rapid growth of their share in the industry from 40 percent of active logging enterprises in 1988 to 70 percent in 1995 (table 1.3). However, as Eba'a Atyi (1998) notes, "large foreign companies continue to dominate the industry because not only do they own most of the production and processing capacity, but they also have a better knowledge of the export market which consumes more than 70 percent of the logging industry production." Thus, while new Cameroonian logging concessions have proliferated, when faced with the difficulties of raising capital through the country's financial institutions, many concessionaires have no choice but to sell or lease their logging rights to well-capitalized foreign-owned concessions with access to external credit markets.[10]

A major policy objective of the government is to increase employment in the sector by increasing the amount of wood processed locally and reducing the export of raw logs. The export value added from processing raw logs into lumber, veneer, and plywood is estimated to have added about 19,000 million CFA francs (US$32 million) to export revenues in 1996, accounting for roughly 12 percent of total export earnings (table 1.4). Approximately 29 percent of the felled timber processed into lumber and plywood enters the internal market, with a fair portion clandestinely exported to neighboring countries. A large portion of the internal market is also supplied from illegally felled timber, which is sawed into lumber by chainsaw operators.[11]

Table 1.3. Evolution in Ownership Patterns in the Cameroon Forest Sector

Year	National ownership		Foreign ownership		Joint venture ownership	
	Number of active enterprises	Portion of concessions exploited	Number of active enterprises	Portion of concessions exploited	Number of active enterprises	Portion of concessions exploited
1988	49	18%	67	77%	3	6%
1996	154	53%	58	46%	8	1%

Source: Adapted from Eba'a Atyi 1998 and *Afrique Agriculture No. 175,* August-September 1990.

Table 1.4. Wood Exports and Export Value Added in Wood Processing, 1996

Product type	Volume exported (m³)	Volume of roundwood equivalents processed (m³)	Total export revenues (million CFA francs)	Avg. export value per m³ of raw log processed (CFA franc/ m³)	Total export value added in trans- formation (million CFA francs)
Raw logs	1,254,407	1,254,407	89,308	71,195	—
Sawn wood	236,340	675,257	60,369	89,401	12,294
Veneer and plywood	35,000	61,403	10,850	176,701	6,478
Total	1,525,747	1,991,067	160,527		18,772

Source: Adapted from Eba'a Atyi 1998.

State Revenues and the Forest Sector

Government fiscal revenues collected from the forest sector are one means of converting Cameroon's forest patrimony into development for the common good. The economic importance attached to the sector by the government of Cameroon and the Bank is indicated by a Policy Framework Paper, which states: *The government expects this sector to contribute to growth and macroeconomic balance.* Both government and donor expectations for an accelerated exploitation of the sector are high. The 1996 Country Assistance Strategy of the World Bank projects a near doubling in forest export revenues from 1996 to 2004 in its baseline scenario. To meet its sectoral revenue and employment goals, it is esti-mated that the 1996 processing capacity would have to double.[12]

Government revenues from the forest sector increased more than five-fold from 1986 to 1995. In 1996 it was estimated that the forest tax collections amounted to 3.3 percent of total fiscal revenues collections, which was much lower than the contribution to total GDP of the forest sector, which was estimated at 8.9 percent. Thus, the importance of forest sector tax collections is still significantly lower than the overall importance of the sector in the economy.

Sustainable Forest Management

The timber industry is often viewed as the principal means of generat-ing value from a forest and is therefore the principal guardian against alternative land uses. If tropical forestry were banned, it is argued, governments would abandon all protection of forest areas. The other non-market and often global values of a forest (existence and option values, carbon sequestration, biodiversity) are yet to be transferred into local incentives, and therefore do not enter into the current calculus of decisionmakers. Thus, in order to preserve the tropical forest patrimony, sustainable management that can guarantee the continued stream of economic benefits from tropical forests must be developed.

One of the principal tools for sustainable management is a forest zoning plan, which Cameroon is in the process of developing. In principle 20 percent of the national territory is to be ultimately included in the permanent forest estate. Of particular importance to the forest sector and the largest component of the permanent forest estate are production forest zones, which are the designated locations for active logging concessions. As of 1996, a total of 62,000 km^2 had been declared as belonging to production forest zones (13 percent of the national territory).

Several projects in Cameroon are investigating sustainable forest management. Among these is the important effort by the French-led *Projet d'Aménagement Pilote Intégré de Dimako*, which is working in the East Province. This is one of several such projects that were developed across Central Africa by the French Ministry of Cooperation and Development following the 1990 Libreville Ministerial Conference on the Valorization and Sustainable Management of Central African Forests. The objective of this project is to *rationally and sustainably exploit* tropical forests, thereby ensuring the conservation of the forest ecosystem, the renewal of forest resources, and the maintenance and optimization of the role of the forest sector in the economy. This exploitation includes the needs of local populations and addresses the improved utilization of forest resources and the reinforcement of technical field services in the forest sector (MINEF 1996).

The most progress has been made on the technical silvicultural aspects of sustainable forest management. The recommended management practices consist of three interdependent parts. The first is to develop knowledge of the resource base, which includes: (1) detailed forest inventories on 1 percent of concession lands in order to establish the tree-diameter distributions; (2) aerial photographs and vegetation studies to establish maps of tree populations; and (3) studies of biodiversity (flora and fauna) to determine fragile areas or particularly rich areas that would require protection. The second is the development and transfer of knowledge for sustainable management of forest resources to the actors involved (state, local population, logging company). The third

dimension is the determination of the parameters necessary for inform-
ing a rational exploitation of the forest. This includes the calculation of
the optimal rotation period based on tree-diameter distributions and
growth rates, determination of the boundaries of the "assietes de coupes,"
and the determination of the minimum exploitable diameters for various
tree species. Through the combination of these three dimensions, the
technical dimensions of sustainable tropical forest management can be
established.

PART II: THE WORLD BANK AND CAMEROON

2

Impact of Bank Interventions on the Management of Forest Resources

Direct Bank Interventions in the Forest Sector

For the 1980–99 period, direct forest sector involvement by the World Bank in Cameroon includes a plantation project in 1982, an ongoing policy-based lending program that started in 1989, and a Global Environment Facility (GEF) biodiversity project approved in 1995. This section of the report examines the effectiveness and sustainability of these operations.

Situation Prior to the Reform

The 1982 forest project aimed to: (1) *expand the contribution* of the forest subsector to the national economy through the establishment of about 11,000 ha of pine and eucalyptus plantations and (2) *enhance the capabilities* of the responsible government agencies to improve reforestation and recover revenue from the forest users.

A self-evaluation conducted in 1992 by Africa Region rated the outcome of this project as poor. Indeed, this project failed to achieve most of its relevant goals efficiently. The two major overall goals of expanding the contribution of the forest subsector to the national economy and enhancing the capabilities of the responsible government agencies to improve reforestation and recover revenue from the forest users appear relevant when considered in their general form. However, their translation into this project may well have been inconsistent with the spirit of the 1978 Forest Policy.[13]

The overall Bank strategy for Cameroon has evolved both in terms of sectoral focus and the instruments used for implementation. Consistent with this shift in overall strategy and starting with the Tropical Forest Action Plan in 1987, the Bank decided to focus on forest policy reforms. Prior to the current legal and regulatory framework (contained in Law No 94/01, which became effective on January 20, 1994, and the Implementation Decree No. 95/531 signed on August 23, 1995), forest sector activities were regulated by the 1981 Forest Law and the 1983 Implementation Decree. These two instruments, however, failed to provide an adequate legal framework for planning land use and integrating forest conservation and production activities with agriculture. The prevailing land tenure regime assigned usufruct rights to anybody who cleared and cultivated land in the state-owned forests that make up most of the dense forest (O'Halloran and Ferrer 1997). This was believed to have encouraged deforestation.

The allocation system for logging concessions used was not transparent and gave the industry incentives to "mine" the forest. There was no requirement for the companies receiving concessions to practice "sustainable" forest exploitation methods. Concessions were awarded for a period of five years and based on mutual agreement between the timber industry and the government authorities. Even though these concession contracts were renewable, it was believed the short period of time might have created incentives for the companies to over-log their concessions. Indeed, logging companies concentrated on a few very valuable species, and due to the extensive nature of the methods involved, they built roads deep inside the forest to reach the sought-after trees (O'Halloran and Ferrer 1997). Thus, the logging activities opened forest areas to individual settlers and bushmeat hunters. In addition, the lack of transparency in the allocation mechanism would breed rent-seeking behavior.

The policy required that 60 percent of production be processed locally. Some people believe that the short period for which concessions were awarded and the high capital cost involved in sawmill investment created incentives for companies to use old and cheaper machinery, leading to inefficiency at a rate of 75 percent wastage (O'Halloran and Ferrer 1997).

With respect to the fiscal system, four major taxes were in effect: (1) A surface area tax at 98 CFA francs per hectare per year; (2) Stumpage tax: a fixed 5 percent of the value of a cubic meter of wood. This value depended on the species and the origin of the log (to account for transport costs). The value of the log, however, was set administratively below market value. In addition, the government relied on the loggers' declarations with respect to the volume and the origin. This information asymmetry led to low revenues from this source; (3) The export tax was a flat tax at 20 percent of an administratively estimated value of the log (rather than the FOB price). This source accounted for about 75 percent of the total amount of taxes collected from the forest sector. This tax created distortions to the extent that processed wood was exempted, thus reinforcing the inefficiencies of the sawmill subsector by providing it a high level of protection; and (4) Finally, there was a specific forest export tax set at 10 percent, aimed at discouraging log export.

In short, the situation prior to the reform was characterized by: (1) lack of adequate legal and planning frameworks; (2) a concession system that encouraged rent-seeking behavior and inefficiency; and (3) a distorted tax system designed to protect an inefficient industry. These are the key issues the Bank sought to address in its policy-based lending.

Content of Current Forest Policy Dialogue

The current forest policy is an outcome of a reform process that started over a decade ago, following the 1987 Tropical Forestry Action Plan. The overall objective of the reform is to improve practices of forest exploitation and management. The new policy intends to correct past non-sustainable practices in natural resource management. It assigns a high priority to the protection of the rich and important biodiversity of the country. The policy has three main components stemming from the recommendations of the 1989 review of the agricultural sector: a legal framework, concession and tax policies, and implementation agencies.

In the 1980s, the forest sector was an integral part of the agricultural sector. Thus the agricultural sector review proposed a forest development strategy designed to help Cameroon increase its share of the world market, increasing the contribution of the forest subsector to employment and national income while rationalizing forest exploitation for industry, fuelwood, and agriculture. The following actions were recommended:

- Development of a *forest utilization* plan to guide forest authorities in the design and implementation of a forest exploitation strategy.
- Reform of the existing *concession system* to provide the private sector with the incentives to regenerate the forest, install modern sawmilling equipment, and reduce uncoordinated agricultural encroachment. In particular, the prevailing five-year limit on concession leases should be extended to at least 20 years.

- As part of a forest *industrialization plan*, invest in training Cameroonian technicians and managers to take an active role in the development of the sector.
- Update the status of existing parks and reserves, identify areas to be included in *protected zones*, and create buffer zones around the protected areas.
- Rationalize the *use of fuelwood* through an overall household energy strategy.
- Reduce the number of *institutions* in charge of forests and clarify their responsibilities.
- Reform the *legislative and fiscal framework*, particularly as it applies to the concession system, wood processing, and the rights and responsibilities of *local communities to manage forest resources.*

How sound is this advice? The forest policy issue confronting Cameroon may be framed as follows. *Forest resources are publicly owned by the government and local communities, while logging resources needed for the exploitation of the forest are in the hands of the private sector, dominated by foreign firms.* The key issue, therefore, is the design of an optimal (enforceable) contract between the public owner of the forest and the private owner of logging resources. There are inherent conflicts of interest in this situation to the extent that the state is (supposedly) interested in an efficient, equitable, and sustainable outcome, while the private sector's interest is limited to its profit-maximizing level of timber harvesting. Under these circumstances, each party would exploit any informational advantage or freedom of action they possess. The contract should try to minimize these potential sources of transaction costs. *Within this framework, we will show later on that, overall, the Bank had a sound idea but could not "make it fly" for lack of a robust implementation strategy.*

Over the years, the Bank tried to push reforms in several areas: zoning and biodiversity protection, concession system, log export regime and local wood processing, and community forestry. Following is a brief summary of the Bank position on each of these points, noting any discrepancy between the point of view of the Bank and that of any other key player.

Zoning and Biodiversity Management

The reform called for the zoning of the tropical forests in southern Cameroon into protected, production, and multiple-use areas. Such zoning is tied to biodiversity management, as it should respect minimum areas for species conservation and genetic biodiversity and allow for corridors between existing reserves and the possibility of transborder protected areas in the context of the Yaoundé Declaration. The Ministry of Forests and Environment approved a planning strategy in June 1999. The plan is considered compatible with sustainable forest management, and its implementation will be monitored under the forest sector floating tranche of the third Structural Adjustment Credit (World Bank 1999: 8).[14]

Concession System

The World Bank recommended a two-tier approach to concession policy. The applicants are first screened based on their technical expertise. Concessions are then awarded based on a sealed-bid, first-price auction whereby the highest bidder wins the concession and pays the price he bid. The object of bidding is the area-based tax (the maximum size of a concession is 200,000 ha). Each year, by the fiscal law, the government set the minimum level of this tax. Table 2.1 shows that the current policy reform has led to a significant increase in the area tax in Cameroon, from 98 CFA francs per hectare and per year in 1994 to (a minimum) 1,500 CFA francs in 1998—about a 1,430 percent increase. There are indications that this tax could yield higher revenue to the government, beyond this minimum of 1,500 CFA francs established by law, as the first competitive offers in 1997 ranged between 1,000 and 4,000 CFA francs per hectare per year.

The idea of auction as an allocation mechanism is sound on efficiency grounds since a bidding mechanism is generally designed to maximize a seller's expected profit (Bulow and Roberts 1989).[15] Indeed, the Bank had hoped that the mechanism of public auction would increase government revenue and reduce corruption in the allocation of the rights involved. In

Table 2.1. Main Forest Taxes in Cameroon

	Fiscal Law 1994/95	Fiscal Law 1995/96	Fiscal Law 1996/97	Fiscal Law 1997/98	Fiscal Law 1998/99
Area tax (CFA franc/ha/year)	98	300	300	1,500-2,500	1,500-2,500
Stumpage (% mercuriale)	5	7	2.5	2.5	2.5
Export tax on logs (%)	25	25	25	17.5	17.5
Export tax on processed wood (%)	15	15	15	12.5	3–4

Source: World Bank (1999: CAS preparatory work).

the particular context of Cameroon, this policy proposal raises equity and institutional issues that have made its implementation quite difficult.

Nationals feel that the system is unfair, as they do not have the same ability to pay as foreigners.[16] This point of view prevailed at the legislative stage. The draft law had provided: "Standing timber shall be sold by public auction to the highest bidder for a non-renewable period of one year." Article 45, section 3 of the (adopted) 1994 Law states: "Standing timber shall be allocated by the minister in charge of forests upon the recommendation made by a competent commission and for a maximum period of one year non-renewable." This outcome also reflects a deepseated resentment by the population toward the plundering of national resources by outsiders. The members of parliament who opposed the public auction based their argument on the predominance of foreign companies in the sector, to conclude that nationals would not stand a chance in such an unbalanced competition. In addition, these members of the parliament were of different political affiliations, including both opposition and the ruling coalition, and of different regional origins.

Sawmill companies oppose the auction system as well because it makes logs more expensive, thereby violating the implicit contract they have with the government for easy access to cheap log inputs. In addition, established logging companies feared that new Asian competitors would take over the forest sector in Cameroon. Hence their interest in limiting entry to the sector.

To deal with these political issues, the Bank did not object to the government practice of discriminatory auctions. In fact, the bidding documents issued in 1996/97 for the attribution of the first 24 concessions explicitly stated that bidding on those concessions was restricted to the enterprises operating in Cameroon. There is evidence that, in the end, some individuals got concessions even though they did not offer the highest bids.[17] This suggests political interference and reveals the true nature of the inequities embedded in the system: the lack of government commitment to abide by the rules and unequal access by agents to the center of decisionmaking.

As far as institutional constraints are concerned, the proposed concession allocation mechanism does not consider the enforcement difficulties surrounding these arrangements and all factors affecting the incentives perceived by the logging companies, particularly foreign companies that operate in a theater much wider than the national scene. In addition, the system is based on a weak informational basis (a source of transaction costs). The relevant attributes of a concession that must be specified and measured in the context of a concession contract are species composition, tree quality, and density (Leffler and Rucker 1991: 1062). The owner of the tropical forest can hardly manipulate these properties. Yet the social value of the outcome (at least in terms of efficiency and ecological sustainability) depends on the care and the diligence with which the logger harvests the timber. This care and diligence are akin to an effort variable, susceptible to shirking, and difficult to observe. In Cameroon, the informational advantage lies with the private sector to the extent that there is no reliable public inventory on the production forest. In addition, the weak capacity of the government (low moral of civil servants following a deep cut in the wage bill, and limited public resources) makes it difficult for the public sector to monitor the operations of a logging company. In Cameroon, the government generally relies on the information provided by logging companies to set and collect fees.

Other institutional difficulties associated with the bidding system in Cameroon stem from the domination of the industry by a handful of firms who, given their opposition to the policy, may collude to subvert the mechanism.[18] In addition, it is not clear that the implementing agencies have the incentives and expertise to ensure a smooth implementation of the system. Karsenty (1998:10) argues that an improvement in the situation may result from the adoption of a Vickery auction mechanism[19] (or second-price auction), whereby the concession would be awarded to the highest bidder at the second-best price. Even though this may improve efficiency under true competition, the equity problem associated with heterogeneity of companies and the weakness of implementing agencies would remain.

There are indications that the government is having a hard time committing to the rules of the game of competitive bidding. In 1996, two large concessions were attributed without auction to two large logging companies. The Bank protested, but to no avail. The same year, 112 *ventes de coupe* (cutting rights) were up for bidding. Bids for the annual area tax ranged from 1,000 to 3,000 CFA francs/ha. There again, not all were granted to the highest bidders. Finally, in August 1997 the government put 26 concessions covering 1.8 million ha up for auction: 190 companies submitted bids, and 16 out of the 26 concessions did not go to the highest bidder (Brunner and Ekoko 1999).

Additional transaction costs may stem from relevant characteristics of logging companies. One may distinguish efficient from less efficient ones; those willing to accept a cut in profit for the sake of preserving the environment and saving some forest for future generations. The forest policymakers in Cameroon are therefore saddled with a typical regulatory problem involving the design of incentive schemes and enforcing mechanisms to deal with the issues created by hidden attributes or actions. The 1998 Country Assistance Strategy (CAS) progress report saw these challenges correctly when it stated the need to provide powerful incentives to concessionaires to manage their concessions in a rational and ecologically sustainable manner and to strengthen the management and monitoring of forest activities. A key point to make in this context is that the creation of the powerful incentives here boils down to the sharing of the socioeconomic rent generated by the "joint venture." The basic principle is to hook the reward of the agent (private sector) to some publicly observable outcome indicator strongly correlated with its effort toward the policy goal.[20] This is a daunting task in an environment characterized by a lack of reliable information.

Additional issues linked to concession policy relate to stumpage fees, the duration of concession tenure (the lease), and management plans. A stumpage fee is a standing-timber-based forest fee. In general, stumpage value is taken to represent the maximum price a buyer would be willing to pay for the standing timber, and thus is an approximation of the price that would prevail in a competitive market (Grut, Gray, and Egli 1991: 10).[21] Estimation of stumpage fees requires a detailed and accurate inventory of the cutting area prior to logging. If there are no additional fees for additional trees cut, stumpage fees provide strong incentives to the concessionaire to harvest all merchantable timber, having paid for the trees. In principle, standing-timber fees also require inspection of logging activities to ensure that they do not go beyond the boundaries of the cutting area. This, in turn, requires a strong administration. In Cameroon, the "stumpage fee" is assessed on the basis of trees felled and not on standing volume. So, except for the area tax, the standing trees are free. In 1996 the stumpage fee was reduced from 7 percent to 2.5 percent of the "*valeur mercuriale*" (a reference price).

As far as lease and management issues are concerned, for the initial interim period of three years following the reform, maximum allowable logging was set at 2,500 ha per year while the concessionaire prepared management plans to be approved by the government. A final contract of 15 years (renewable) would then be based on an approved management plan. Few management plans have been completed and none has been approved to date (World Bank 1999: 4). Some observers point to this as the clearest evidence of the failure of forest policy reforms in

Cameroon. Companies have been getting their timber mostly from these *"vente de coupes"* with no environmental "strings" attached. This is happening 10 years after the initiation of the reforms.

Log Export Regime and Local Wood Processing

The government of Cameroon believes that the country would benefit more from its natural resources by processing them locally. That is why the 1981 policy advocated that 60 percent of the timber production be processed locally. This target was maintained in the bill submitted to the National Assembly. But during the debate, the majority of members (including the ruling coalition) insisted on a complete ban on log exports, advocating the local processing of all logs prior to export. There are reports that the government made a big push to have its allies in the assembly toe the party line. A compromise position was found: 70 percent became the new requirement[22] until 1999; after that, the total ban would take effect.

On June 30, 1999, a logging ban was imposed on the rare hardwoods including iroko, moabi, and bubinga. Conditions under which other species of wood could be exported remain to be spelled out by the government. This is a compromise solution. Initially, the government intended to impose a total ban. This was opposed by the industry and the World Bank. The Bank seems to have accepted the current partial ban. A total export ban is considered by some as the strongest incentive that can be provided for domestic processing. In principle it is simpler to implement and easier to monitor. It is a clear-cut policy. It is, however, inflexible. It cannot be easily adjusted as the conditions change or as the local industry matures. It can only be removed or reinstated (Grut, Gray, and Egli 1991).[23] This partial ban is desirable as far as the moabi tree is concerned; in fact, it does not go far enough. Given the nutritional and medicinal importance of the moabi tree, it would be desirable to ban logging it totally. *In fact, management plans must be devised in collaboration with local people to ban logging of similar trees.*

It appears that the government has not yet reached a definite (irreversible) position on the log export ban. On August 31, 1999, the President of the Republic signed an ordinance[24] that seems to reimpose a total log export ban. In essence, the ordinance imposes a total export ban starting from December 31, 1999. Up to that point, the industry could process locally up to 70 percent of its log production. While seemingly upholding the provisions of the 1994 forest law pertaining to the export ban, the ordinance also opens a big loophole by stating that, once the ban becomes effective, log exports would continue for the promotion of certain species and that such exports would be subject to a surtax. Given the weak institutional environment prevailing in Cameroon, it is easy to imagine cases of mislabeling and other fraudulent activities to subvert the ban.

Besides the log ban, log export policy in Cameroon involves an export tax and a domestic processing requirement. A log export tax, in principle, will divert logs to the domestic market, thereby lowering domestic log prices. High export taxes on logs combined with low or zero export taxes on processed products are supposed to provide a strong incentive for domestic processing as this tends to lower log prices below world market prices and provide domestic processors with cheap log inputs (Grut, Gray, and Egli 1991:30). Low-priced domestic logs will induce inefficiency in utilization in the forest and in processing plants. This may mean that some species will be left in the forest.[25]

The structure of the export tax contained in table 2.1 also reveals a preferential treatment of the processing industry compared to log exports. This is an outcome of the pressure the private sector brought to bear on the government in September 1996 to win a significant reduction of the export tax rate applicable to log inputs into local processing.

A key objective of the reform proposed by the Bank in Cameroon is to move the tax base from downstream to upstream of the production process. This is being accomplished to some extent through the area tax. In July 1995, in the context of the first Financial Law following adoption of the Forest Law, the government tried to make the local wood processing industry pay an ad valorem export tax of 25 percent on their log inputs. The private sector fought back with intense lobbying and a threat (issued on September 9, 1995) to close their operations (on October 11, 1995) if the Ministry of Finance did not reduce the tax. The administration took the threat seriously enough to back down and reduce the tax to the satisfaction of the industry (Carret 1998: 56). The government tried again within the 1996/1997 Financial Law to subject log inputs to the same export tax as exported logs. On September 7, 1996, processing firms responded by laying off their employees. Two days later, the government agreed to bring the export tax rate to a much lower level, 3–4 percent (Carret 1998: 69).

The World Bank generally supports the objective of promoting a local industry as it figures among the recommendations of the 1989 agricultural sector review. The Bank opposes the way the government has been going about it. It relied on the wrong instrument by protecting the existing industry (export ban and preferential treatment). The work force of the logging industry is structured as follows: 3 percent executives, 5 percent supervisors, and 92 percent laborers (unskilled workers; Eba'a Atyi 1998:14). An equitable industrialization policy that ensures international competitiveness of the industry would require investing in training of Cameroonian technicians and managers to take active part in the development of the sector. This proposal (by the Bank) has been on the table for ten years. *Capacity and private institutional development have not been pursued under the current policy.*

Community Forestry

The 1994 law addressed the recommendation by the World Bank that local communities be actively involved in the management of forest resources. However, the rights and responsibilities of the communities have not yet been specified. There is no clear mechanism to ensure that the local elite does not capture the intended benefits. Furthermore, the receipts from the area tax were supposed to be shared with the local communities, but this has not yet been implemented fully and transparently.

Delivery Mechanisms

The forest policy advice of the Bank to the government of Cameroon was formulated in the context of the 1987 Tropical Forestry Action Plan, the 1989 review of the agricultural sector, and subsequent Country Assistance Strategies. The preferred delivery instrument is a policy-based lending program that started in 1989. We review both the CASs and the adjustment program.

The Strategic Framework

The 1994 CAS for Cameroon was prepared along with an Economic Recovery Credit in the amount of US$75 million, on standard IDA terms with a maturity of 40 years to support major economic reforms and provide general balance of payment support. The CAS identified three major development challenges: an overextended and inefficient public sector; a deteriorating human and physical capital base and declining productive capacity; and increasing poverty.

Given the development constraints identified above, the 1994 CAS set out to help the government achieve three objectives: (1) improve the performance of the public sector through downsizing and reorganizing of the civil service, institutional reforms of the public procurement system, and rational use and management of natural resources; (2) improve productive capacity through reforms of the legal and regulatory framework, improved macroeconomic management, maintenance and development of basic infrastructure, development of human resources, and strengthening the financial sector; and (3) improve the delivery of social services to low-income groups.

With particular reference to the forest sector, the strategy seeks to ensure a more efficient and transparent management of Cameroon's forest resources by helping the government formulate and implement a new Forest Law. In fact, the implementation decree for the new law was made a condition of effectiveness of the 1994 Economic Recovery Credit, backed by the CAS. The Bank is insisting on the transparency of the procedures for awarding logging concessions, on reducing the role of public institutions, and increasing the participation of local people. The Bank also recommends the cancellation of the ban on log exports.

Within the context of the chosen evaluative framework (see box 2.1), it is clear that both the overall strategy and the forest strategy are relevant given the development constraints identified. The objectives are consistent with the fundamental Bank mission of poverty alleviation and environmental sustainability.

It is not at all clear that both strategies are fully owned by the key players and stakeholders. There is no indication that the CAS was prepared in a participatory manner. The document notes that dialogue with the country has been rather difficult and this has precluded groundwork for sustainable development. The decision of the government to go along with the reform package associated with the 1994 CFA franc devaluation was interpreted as a significant improvement in policy dialogue. As far as the formulation of the forest policy is concerned, there are reports that the views of local people were not sought or taken into consideration.

The 1994 CAS document lists activities by the IFC and MIGA, with no indication of the existence of a coordinating mechanism to ensure that internal synergy is exploited. As far as cooperation with other donors is concerned, the strategy explicitly includes informal meetings of donors in Washington and Europe and informal monthly meetings at the World Bank Country Office in Yaoundé. It also advocates the maintenance of meeting of the sector coordination groups.

The 1994 CAS correctly identified institutional weaknesses in the public sector as a serious constraint to development, and proposed activities such as a public administration reform project and a private sector development operation. However, the program does not include activities to develop the capacity of the civil society. The strategy also lacks a proper information policy.

Box 2.1. A Simple Framework for Assessing the Treatment of Forest Sector Issues in a CAS

Administratively, a CAS is the main vehicle the Bank Group's Board of Executive Directors uses to review the assistance strategy for borrowers. The document contains a plan of action designed to help a country achieve its development objectives within the context of a political economy. It determines the overall level and structure of Bank activity depending on country circumstances as determined by performance and risk.

How satisfactory the treatment of forest sector issues is in a forest-rich borrower such as Cameroon depends on several considerations. In this context one may view the CAS in terms of two basic components: (1) the overall strategy and (2) a forest sector strategy embedded in the overall approach.

Any evaluative framework must answer the following three questions: What are the valuable aspects of the object of evaluation? How valuable are they? How can one formulate an aggregate judgment? We base our simple evaluative framework on OED's doctrine of development effectiveness. The OED framework uses the following criteria: relevance, efficacy, efficiency, institutional development, and sustainability. To invoke this approach in the context of a strategy we consider some key determining factors of development effectiveness and sustainability: relevance, ownership, synergy, institutional development, and information.

The selection of the above factors also stems from the view that development achievement by a client country is an outcome of a strategic interaction between the country, the Bank, and other stakeholders. Such an outcome is successful when there is compatibility among the objectives of key players. It is therefore accepted that the effectiveness of development assistance requires, in addition to the relevance of its objectives, a strong country ownership and broad stakeholder involvement. In addition, efficiency in implementation requires working partnerships based on comparative advantage to exploit synergies. Successful institutional development makes sustainability of outcomes more likely. Finally, a good strategy is supported by a good information policy aimed at minimizing transaction costs among socioeconomic agents. Good and relevant information is necessary throughout the strategy cycle. The formulation needs an adequate diagnosis of the problem and its determinants; proper information dissemination facilitates implementation, and a good monitoring and evaluation system helps in outcome assessment. Overall, a good information policy seeks to achieve equal access to relevant information by all stakeholders or socioeconomic agents.

For a given CAS, the five dimensions of the framework are scored as follows. Each of the aspects receives two points if treated satisfactorily for both the overall CAS and the embedded forest sector strategy, one point if the treatment is satisfactory for one of the two core elements, and zero points if the treatment is unsatisfactory. The aggregation rule is a simple summation. The overall grade thus ranges from zero to ten. We require at least an eight for a good CAS.

This CAS is rated a 5 out of the 10 possible points. A disaggregated rating is presented in table 2.2. This rating reveals that the strength of the CAS process for Cameroon lies in the relevance of the objectives, the concern with institutional and governance issues. The process is weakest in terms of mechanisms to promote full ownership and to deal with information deficiencies.

The 1996 CAS is basically an extension of the 1994 CAS with an effort at improving consensus building and reform sequencing. It does not, therefore, deserve a stand-alone score. It stated the following as the overall objective: "to help Cameroon recover lost ground in economic and social development as soon as possible, while further restructuring its economy to remain competitive in a rapidly changing world." This objective is consistent with that of the 1994 CAS. The basic difference between the two strategies is in the sequencing of the implementation of reforms, the attempt by the 1996 CAS to (1) link more closely actual performance and the availability of funding and (2) broaden consensus around the reform agenda while stimulating debate within the government and between the government, the private sector, and the NGOs. *This validates, to some extent, the low score given to the 1994 CAS on ownership.*

A Country Assistance Strategy Progress Report presented to the Board in March 1998 made the following assessment in the areas of poverty alleviation, natural resources and biodiversity, and governance. The report noted the lack of transparency in government operations due to endemic corruption throughout public services and the judiciary (World Bank 1998a: 5). To deal with this issue, a program is under development with the support of UNDP to (1) strengthen public administration; (2) achieve effective decentralization and build local political capacity; (3) reform the judiciary; and (4) enhance citizen and civil society participation in the formulation and implementation of public policy. The report also states that the adjustment program will continue to support the establishment and application of a transparent system of concession allocation.

Table 2.2. Rating the 1994 CAS for Cameroon

Field	Score
Relevance	2.0
Ownership	0.5
Partnership (synergy)	1.0
Institutional development	1.5
Information	0.0
Total	5.0

Among the achievements listed for the management of natural resources and biodiversity, the progress report listed the revision of the Forest Law and the completion of a National Environmental Action

Plan. The remaining challenges include: (1) the institutionalization of a competitive system of allocating forest concessions; (2) providing powerful incentives to concessionaires to manage their concessions in a rational and ecologically sustainable manner; (3) strengthening the management and monitoring of forest activities; (4) enhancing the effectiveness and fairness of forestry taxation; and (5) devising and implementing an adequate mechanism for the sharing of tax revenue from forestry with the local communities.

The report notes that poverty alleviation is the area where very little has been accomplished since the January 1996 CAS. A series of standard recommendations are made, including the advocacy of a greater participatory approach to poverty reduction.

The Evolution of the Structural Adjustment Program

Cameroon has been under structural adjustment since 1989 when the first Structural Adjustment Loan (SAL) for US$150 million was approved (Loan 3089-CM). The loan was approved on June 14, 1989, and became effective on November 28. The broad objectives of the loan were to help the country redress the severe decline in GDP through internal adjustment measures to enhance competitiveness, reorienting the role of the State away from direct market intervention, toward the design and implementation of policies that are more supportive of the private sector. The specific components for reform cover the following areas: (1) public resource management; (2) civil service; (3) public enterprises; (4) financial sector; (5) liberalization of trade and prices; and (6) the development of productive sectors including agriculture, forests, and manufacturing. Thus, forest sector issues made their appearance in this initial SAL before the promulgation of the 1991 Forest Strategy. The Country Team must have been forward-looking in the design of the program. Another possibility is that they were influenced by the development of the Bank's 1991 Forest Strategy.

In the case of forest sector, the adjustment sought to implement the recommendations made in the 1989 Agricultural Sector Review. In particular, the Memorandum and Recommendation of the President stated that the reform will: (1) involve the revision of the forestry code to provide incentives to the logging companies to invest in maintaining and controlling access to their areas of operation; (2) revise the fiscal system in a way that provides incentives for the exploitation of little-known species and for better wood processing; and (3) revise the forest code in order to provide a legal framework for agroforestry, and the integration of local population in the protection of national parks and reserves.

The disbursement of the loan was divided in three tranches of US$50 million each, based on 19 conditions. Four conditions were to be fulfilled before loan effectiveness. For the disbursement of the second tranche, seven conditions were to be met by March 31, 1990. The remaining eight conditions, including the revision of the forest code, were associated with the third tranche. These conditions were to be met by December 31, 1990. The government made progress in complying with many of the specific conditions attached to the third tranche, but could not meet the general condition associated with a viable macroeconomic framework. It then became apparent that the SAL condition of a satisfactory macroeconomic framework could not be met without a devaluation of the currency. This message was carried to the government by the supervision mission of April 1993.

Over time, poverty increased as the economy continued to decline. By the 1994 devaluation, Cameroon had lost its creditworthiness for borrowing on IBRD terms and was declared eligible for IDA funds. As a consequence, the third tranche of the SAL was transformed into a single-tranche Structural Adjustment Credit (SAC) in the context of a Stabilization and Adjustment Program supported by an 18-month Standby Arrangement with the IMF (approved on March 14, 1994). The SAC was released even though the condition pertaining to the forest code had not been met. The Bank argued that the general condition for maintaining a satisfactory macroeconomic framework had been met, and granted a waiver for the forest condition. An Economic Recovery Credit (ERC) for US$75 million was approved in June 1994 as part of an emergency package to the CFA franc countries following the devaluation. In this context, the Bank insisted that the government adopt implementation decrees satisfactory to IDA by end-June 1994, and resubmit a revised code to Parliament by November 1994. The first requirement was made a condition for effectiveness.

The outcomes of both the SAL and the SAC were rated unsatisfactory. The outcome of the ERC is considered broadly satisfactory. Sustainability is rated unlikely for all three operations.

The main objective of the 1996 SAC was to improve public finance management. Even though it did not include an explicit conditionality on the forest sector, it did reiterate the position of the Bank with respect to forestry taxation. The report stated that for the government to achieve sustainable forest exploitation, promote an efficient wood processing industry, and secure an adequate level of fiscal revenues, it is necessary to: (1) improve the collection of concession and stumpage fees; (2) increase both the stumpage and the minimum concession fees; (3) differentiate the rate of stumpage fees by species (charging higher fees for higher-value species); and (4) reduce export taxes.

There was no Structural Adjustment Loan to Cameroon between 1995 and 1998. A review of past adjustment experience in Cameroon by the design team of the Third Structural Adjustment Credit (June 1998) noted the following lessons: (1) it is important to have a credible macroeconomic framework; (2) poverty has to be brought to center stage of the reform agenda; (3) enhancing growth while protecting the environment entails deep structural reforms in four areas: transport, privatization, financial sector, and forest sector; (4) the reward should be linked to outcomes rather than inputs or promises; and (5) it is essential that the government have both ownership of the reform program and commitment to the implementation of the reforms. This is a clear indication that the Bank has been trying to follow a multisectoral approach in Cameroon.

The Third Structural Adjustment Program sought to address (1) the absence of appropriate incentives and institutions for endorsing and monitoring management plans; (2) lack of procedures for the determination of community forests and the involvement of local populations in their management; (3) a distorted fiscal system, with lower export taxes applied to logs processed locally; (4) the log export ban to be imposed in 1999; and (5) noncompliance by the authorities with the law with respect to the concession award system.

To deal with the above constraints, the program adopted a strategy to (1) implement an institutional framework with appropriate incentives for the approval and compliance with management plans by logging companies, including monitoring and enforcement mechanisms; (2) consolidate tax reforms; (3) review the desirability of the log export ban; and (4) improve the allocation mechanism for cutting rights and concessions and reduce political interference. The implementation of this strategy is linked to the structure of the conditionality in such a way that the forestry component of the credit is made contingent upon agreed actions prior to Board presentation, release of the second tranche, and release of the forest sector floating tranche.

Interventions by the World Bank in the Agricultural Sector

From its first mission to Cameroon in 1964 until the late 1980s, the World Bank viewed Cameroon as a country with vast unexploited and underexploited natural resources. Over this period, the Bank's proposed strategy for Cameroon was the intensification of agricultural production and simultaneous area expansion to bring idle resources into production to facilitate economic growth. The Bank's priorities in policy dialogue, project lending, and conditionality at the close of the 1980s were: (1) strengthening of government organizational capacity to formulate and implement policies for intensification and expansion of agriculture and forests; (2) massive development of forest resources, particularly in the east and the south; (3) increased efficiency in the smallholder sector; (4) extensive upgrading and expansion of the plantation subsector; and (5) large-scale expansion of livestock production on underused savanna lands.

Agricultural projects declined from seven in the eight years leading up to the Bank's 1991 Forest Strategy to only two in the eight years since. These figures compare with 6 projects from 1967 to 1975 and 11 from 1976 to 1983. The major emphasis in the early years was on the development of the industrial tree crop plantations in the southwest. Later, in the 1970s and 1980s, several projects addressing intensification were initiated with relevance to the moist forest resources. These included a cocoa project, several integrated development projects, and a program on agricultural research. Since the 1991 Forest Strategy, only the training and visit extension project has directly addressed the intensification of smallholder agriculture, while a food security project has had potential for indirect effects. Of more importance in the post-1991 period has been the effect of Bank policy recommendations and conditionalities in the context of structural adjustment, including the breakup of the export marketing board for cocoa and coffee, price liberalization, input market liberalization, and reform of the cooperative law.

Negative Impacts of Policy Sequencing on Traditional Export Crop Sectors

The World Bank played a significant role in the liberalization of cocoa and coffee marketing and the breakup of the Office National de Commercialisation des Produits de Base (ONCPB) in 1991. This was accomplished through loan conditionalities during disbursement of Cameroon's first structural adjustment loan of US$150 million. It was also incessant in calling for devaluation, which finally transpired in January 1994. Unfortunately, because of the political economy of liberalization and the devaluation (see Annex G), the sequencing of these policy reforms in the cocoa and coffee sectors was less than ideal for producers. The removal of the 50 percent subsidy on fertilizers was the first policy reform implemented. Subsidies were gradually phased out from 1988 to 1992. [26] This was followed in 1990 by a 40 percent cut in the official producer prices of coffee and cocoa by the ONCPB, which was unable to maintain stable producer prices in the face of the overvalued CFA franc and following the depletion of its reserves by a fiscally strapped government.

Producers responded by significantly curtailing resources allocated to cocoa and coffee agroforests, including fertilizers.[27] With world coffee and cocoa prices at historical lows, state-administered panterritorial pricing was phased out in 1992 for coffee, but not until 1995 for cocoa. Price liberalization at a time of historically low world prices and an overvalued exchange rate resulted in farm gate prices that were less than half their nominal 1988 levels. At these prices, many farmers did not even harvest their coffee. At the same time, the liberalization of fertilizer markets and the de facto liberalization of pesticide markets not only increased producer costs but also reduced availability due to an inadequately developed private sector. Fungicide control of cocoa blackpod disease fell dramatically from over 30 million packets of fungicide distributed *gratis* by the state in the mid-1980s to less than 3 million purchased from private suppliers in 1993 (Varlet and Berry 1997). Finally, when devaluation came along in 1994 and doubled nominal producer price, the supply response was muted by a decline in the productive capacity of cocoa and coffee plantations, which had been, at best, minimally maintained under the policy regime from 1989 to 1994.

The best sequence of policy reform would have been to first devalue the CFA franc, second, liberalize cocoa and coffee marketing, and third, phase out fertilizer subsidies. Côte d'Ivoire provides an interesting contrast to Cameroon. The Ivorian government was able to maintain research and extension support for cocoa during the crisis in world markets and the overvaluation of the CFA franc. Since devaluation in 1994, Côte d'Ivoire has increased its production of cocoa by more than 600,000 tons, whereas Cameroon has not seen any significant increase. In sum, the unintentional effect of the structural adjustment conditionalities was to seriously handicap Cameroon's smallholder export sector by a significant depreciation of farmers' tree stock. When prices in the mid- and late-1990s rose, farmers were unable to respond robustly.

Estimates of the impact of this policy-led extensification on the cocoa sector within the Forest Margins Benchmark include a 40 to 50 percent decline in producer revenues and an estimated decline in the carbon stocks ranging from 0.21 percent to 2.14 percent that occurred when producers retracted resources from cocoa agroforests and increased the cultivation of food crops (Gockowski et al. 1998). Had these policy shocks been permanent, it is estimated that a total decline of 4 to 17 million tons of carbon in the four administrative divisions would have occurred in the Forest Margins Benchmark alone. With the possibility of carbon trading for over US$20 per ton, this shows how the environmental costs of policy actions can be substantial, even in a relatively small area such as the Forest Margins Benchmark.

Instead of asking the question of where Cameroon's future comparative advantage was likely to lie in the late 1980s and 1990s, when commodity prices were at historical lows, and perhaps deciding that the economically important coffee and cocoa sectors of smallholder producers might have required support to help them adjust to a temporary shock in world commodity markets, the Bank instead pushed for its standard liberalization package. As a result, the country witnessed a serious extensification of its coffee and cocoa agroforestry systems.

Today, the economic environment for intensifying these systems is much better. In the 1970s and 1980s, implicit producer taxation occurred through panterritorial pricing at levels averaging 50 percent below FOB prices, and marketing margins were fixed by the marketing board, with monopsony purchasing rights assigned to licensed buyers. The impact on the environment and the welfare of smallholders could have been minimized if the political will of the government to support these subsectors had been stronger and if the Bank had not been so stringent in some of its policy recommendations. While the cocoa sector is finally starting to respond positively to world price signals and the new liberal economic environment, output from the coffee industry is still only half of what it was before liberalization in the mid-1980s.

Rehabilitation of the Cocoa Sector in the Center and South Provinces

The Société de Développement du Cacao (SODECAO) was created as part of a project funded by France and the World Bank to intensify cocoa production in the Center and South Provinces. The program was initiated in 1975, for seven years, in hopes of building on the optimistic technical results obtained by the French and USAID in several pilot projects conducted in the late 1960s and early 1970s. Emphasis was placed on the multiplication and diffusion of "hybrids," on the control of cocoa blackpod disease,[28] and on extension. The initial project appraisal assumed that the relatively low yields of existing cocoa agroforests could be increased from 300 kg/ha to 500 kg/ha, mainly by providing farmers with the necessary fungicides and sprayers so that the research-recommended levels of fungicides (12 sprayings including application during the dry season) would result. The new plantings of "hybrids" were expected to produce 1,200 kg/ha at maturity (years 12 to 25). These results were never approached for numerous reasons, and at the completion of the project the economic rate of return was negative. The project was beset by organizational and institutional problems and technical deficiencies. Perhaps chief among them was the failure to recognize the constraints of a limited smallholder labor supply and underdeveloped labor market, which prevented the implementation of the recommended fungicide spraying program on a large scale.

In the mid-1980s the Bank again entered discussions for rehabilitating the cocoa sector in southern Cameroon. A large-scale US$285 million project with a Bank loan of US$103 million was approved and made effective in 1988 with cofinancing by the Caisse Centrale de Coopération Economique (now the Caisse Française de Développement). The project intended to increase cocoa production in the Center and South by 60 percent by again targeting new plantings and yield increases. The project was to be implemented by SODECAO and was intended to improve producer incentives, reorganize the marketing system, establish a private medium-size plantation program, and strengthen the Ministry of Agriculture's capability to implement the cocoa strategy and to monitor its execution. With the continuing economic crisis in Cameroon, it became clear that government would be unable to maintain its financial commitments to the project. In 1991, project objectives and the means to achieve them were significantly altered, amended, and resubmitted, after which time the loan was reduced to US$62 million. The major reorientation in project means was a shift toward liberalization of cocoa marketing, reform of other policy distortions and public sector institutions, the substitution of independent farmer groups, which flourished after the 1991 reform of the cooperative law, in lieu of the government-controlled cooperatives, and the gradual withdrawal of SODECAO from infrastructure construction and input supply activities. The loan was closed in 1995, at which time a balance of US$29 million was canceled. The project resulted in no measurable increase of cocoa production or yields. Productivity decreased in the project area as farmer incentives drastically declined, mainly due to inappropriate sequencing of policy reforms under the project and the concurrent structural adjustment program. In sum, the US$74 million spent on the project over the eight years of implementation represents a cost of approximately 55 CFA francs per kilogram of cocoa produced.[29] This cost represents more than 20 percent of the official producer price during the project implementation period. In retrospect, a simple Pigouvian support of the producer cocoa price to alleviate the economic distortion caused by the overvalued exchange rate would have resulted in much greater impact than this project, which effectively accomplished nothing; through the disruption wrought in input supply and extension services, it actually led to lower producer incentives.

Agricultural Research

Agricultural research was most recently directly supported by the National Agricultural Research Project, which was effective from 1988 until 1993. Donor funding was US$17.8 million by the World Bank, with GTZ funding of US$3.3 million and ODA funding of US$2.8 million. The project's strategy, framed within the overall sector strategy, was to increase the productivity and incomes of the small-scale agricultural producers through technology creation and transfer. Specific objectives were to: (1) improve research programming of the *Institut de Recherche Agronomique* (IRA) and the *Institut de Recherche Zootechnique et Veterinaire* (IRZV); and (2) improve the technical and operational capabilities of those institutions by establishing a system of farming research management.

Agricultural research, which has recently been reorganized, is still not operating up to its potential. IRA and IRZV were melded into a single agricultural research institute—IRAD—in 1998 with an operational loan from the African Development Bank for US$5 million over five years, effective in July 1999. There is also a need for the institute to adapt the innovations in civil society, which have included the rapid development of farmer organizations in the last eight years. The voice of these organizations needs to be heard and incorporated into the research planning and priority setting process. Researchers are also hampered by a lack of communication infrastructure (most researchers do not have access to the Internet) and the research library (whose improvement was one of objectives of this project), which has no budget for scientific journals.

Programme National de Vulgarisation Agricole (PNVA)

This training and visit agricultural extension program began as a pilot project in six provinces including the Southwest, Littoral, and East Province among the five forest provinces. The US$31 million project began its operational phase in 1991; a second phase, with national coverage, began in 1997. In 1997, the project employed 2,394 personnel, 69 percent of whom were field extensionists, while 11 percent were regional technical specialists, and 20 percent had supervisory roles. The project also has an adaptive research component led by IRAD involving on-farm adaptive trials of promising technologies and monthly technical training of extension workers.

With the increased mobility of extension agents following the distribution of project motorcycles, a substantially higher proportion of farmers has been reached. Another encouraging aspect of the program has been the willingness of the service to work with the private sector. In the Forest Margins Benchmark, for instance, a private supplier of fertilizers is furnishing plantain growers with fertilizer recommendations through extension agents. Plantains, the most important commercial food crop in the forest zone of Cameroon, are typically grown in a very extensive fashion and are often targeted to long-fallow fields, in part because of a high nutrient demand. Intensification of plantain-producing systems should be one of the focal points for food crop research in the forest zone. Subregional delegates of agriculture (SDAs) in the Center Province were asked what they appreciated most about the new program vis-à-vis the old system. There was near unanimous agreement that the increased field exposure of supervisory staff and contact with the rural citizenry was increasing the relevancy of the message. The contacts between farmer groups and the extension agent in the farmers' fields also contributed to a more relevant message. Finally, it was noted that the project was leading to more accountability of both extension agents and their supervisors.

Among the problems noted by the national coordinator of PNVA are organizational insufficiencies such as a lack of qualified technicians, especially women, to serve extension demands and the arbitrary relocations of extension agents (A. Nyaga, personal communication).[30] The dislocation of extension agents has a negative impact on the on-farm research program when these agents, used by researchers to implement trials, leave unexpectedly without any handing-over of their program to the new agent (B. Nguigoumou, personal communication). In the SDA interviews, several other issues were raised. For instance, in a major cocoa growing area, a farmer's organization had undertaken the insecticidal treatment of farmers' cocoa plantations, which—according to the subdivisional delegate of agriculture—had undermined the confidence of farmers in the extension service.[31] The maintenance of motorcycles after only one year in operation was also already becoming a major issue, and some of the SDAs indicated that the motivation of extension agents was diminished by salary arrears of up to eight months.

One of the biggest concerns of many observers of the program is the lack of technological solutions generated by research for addressing farmers' problems. This is related to the problems that agricultural research has faced in Cameroon over the last 15 years. With five-year lags before most research begins to have a significant effect on productivity, even an immediate infusion of research resources is unlikely to lead to substantial innovations before project closure in 2001.

Interventions in the Transport Sector

During the 1975–86 period, the government transport strategy consisted primarily of large investments to address increasing sector needs. These large investments were made possible in part by the generous amount of donor funding that was available during the period. Currently, the strategy for the transport sector focuses on alleviating several constraints that have contributed to poor performance of the transport sector in Cameroon:

- The road sector, particularly road maintenance, has received insufficient attention from the government, even though this sector constitutes the most important transport mode in Cameroon and road transport activities account for 90 percent of both overall passenger and freight transport and is spread all over the country. In fact, even when different transport modes compete on the same routes, road transport attracts an estimated 70 percent of the transport demand through lower tariffs and greater efficiency. Basic road maintenance expenditures have been significantly reduced over the years since 1986, due primarily to declining government revenues. Thus, government funding for basic road maintenance was cut to less than a third (from 25 billion CFA francs in 1986 to 7.5 billion in 1991), resulting in an annual shortfall of 39.5 billion CFA francs (US$144 million) for road maintenance. This underfunding of road maintenance has led to a severe deterioration of the earth road network and decapitalization of assets.
- Public enterprises have been poorly managed and accumulated huge annual deficits leading to bankruptcy.
- Development projects were proposed without appropriate economic analysis, intersectoral balance, and consideration of the country's limited financial and implementation capacity.

The new transport strategy, supported by the World Bank, seeks to address key institutional and sector policy constraints that have contributed to the poor performance of the sector in the areas of planning, policy formulation, maintenance operations, and equipment handling and management.

The objectives of the new transport sector strategy are to:

- Strengthen road maintenance capacity through the development of a new road maintenance policy.
- Improve road transport fiscal policy and cost recovery.
- Restructure public enterprises.
- Improve institutional capacity and strengthen transport sector management.
- Improve transit transport facilitation and the corridor competitiveness.

The road maintenance strategy proposes to formulate maintenance programs that: (1) take into account existing budgetary constraints of the government as well as relative levels of service and traffic volumes of paved and earth roads networks; (2) disengage the State from maintenance work on all paved roads and the majority of rural roads by contracting out maintenance services to private enterprises; (3) use labor-intensive methods and local materials, whenever possible; (4) protect the environment in areas along roads to be maintained by favoring non-mechanized methods whenever possible and incorporating environmental safeguards and the use of local materials; and (5) clarify and coordinate responsibilities for road maintenance among the Ministry of Public Works, the Ministry of Urban Development and Housing, and the local authorities, in order to optimize the use of scarce fiscal resources and improve consistency in the development of interurban, urban, and rural road networks.

Road maintenance activities will aim at preventing further loss of assets and future costly rehabilitation works. They are expected to have a significant positive impact on the country's economy by reducing transport costs and generating local employment.

The 10 public enterprises operating in the transport sector will be restructured and more commercially operated by widening ownership, encouraging staff buyouts, and increasing transparency and accountability in management decisions. If these strategies fail, they will be fully privatized.

Competition between transport modes and among operators within each transport mode will be promoted in order to reduce transport costs.

Environmental Issues Raised by Chad–Cameroon Pipeline

The Chad–Cameroon Petroleum Development and Pipeline Project is expected to generate substantial revenues for both Cameroon and Chad.

For Cameroon, the project is to generate an estimated US$500 million in transit fees, taxes, and dividends from the transport of oil from Chad to the Atlantic Ocean.

The project's physical components include:

- Drilling approximately 300 wells in the Doba fields in southwestern Chad, which hold about 900 million barrels of oil and construction of associated facilities and infrastructure
- Constructing an export system that consists of a 30-inch-wide, 1,050-kilometer (650-mile) -long buried pipeline from the Doba fields to Cameroon's Atlantic coast
- Installing a marine (offshore) terminal and marine pipeline facilities in Cameroon.

The pipeline will stretch through a variety of ecological landscapes and to the Atlantic Ocean. In addition, the project will displace a small number of families (between 60 and 150 households) from their farmlands and bring about their involuntary resettlement around the oil field in the Doba basin. The project's construction activities will also involve removal of vegetative cover and shade canopy, thereby increasing soil surface temperature, decreasing moisture content, killing soil organisms, and increasing the potential for erosion in the oil field development and along the pipeline. It may also lead to increased peak flows and sediment loads of small tributaries of drainage and reduced water quality caused by increased wastewater discharge.

The Bank states that it is very sensitive to these concerns and its decision to support the project is dependent upon three major factors: the project's potential to reduce poverty and increase public investments in health, education, and basic infrastructure; satisfactory internal reviews of environmental assessments, management, and resettlement plans submitted by each government to the Bank; and agreement with each government and the private sponsors (Exxon, with equity of 40 percent; Petromas, 35 percent; and Chevron, 25 percent) on the implementation of agreed-upon plans to mitigate any damage to the environment.

In response to the Bank's requests, the sponsors have hired a reputable firm to prepare environmental assessment plans for the project and to assist in developing a route that is environmentally acceptable. The governments of Cameroon and Chad have submitted their Environmental Assessment Reports and Environmental Management Plans to the Bank. A Resettlement Plan has also been submitted. All these documents must be reviewed and approved internally within the Bank before the credit can be processed.

The project proposes to mitigate any damage to the environment through the following environmental measures:

- Disturbances to forest and riverine vegetation during construction will be minimized by controlling unauthorized use of the pipeline route during construction.
- Increased peak flows and sediment loads of small tributaries of drainage will be corrected by draining surface runoff to more than one tributary.
- Disturbances to existing local supply wells will be reduced by staggering the water supply wells built under the project.
- Sanitary wastewater discharges from the project will be treated in compliance with the World Bank effluent guidelines to correct water quality problems caused by the project.

This project represents the largest single private investment in Sub-Saharan Africa, and is also the most controversial. Most of the NGO community (led by the Rainforest Action Network) argues that oil projects in poor countries governed by corrupt regimes do not reduce poverty. In addition, such governments lack the political will and the capacity to implement regulatory measures capable of mitigating the risks of environmental and social destruction. They cite the case of Nigeria to bolster their point. Thus, according to these NGOs, the most likely outcome of this project is exacerbation of corruption and human rights violations.

Given the records of both governments on these issues, it is very hard to dismiss the NGO arguments. In fact, the Bank recognizes that an important risk here relates to the weak capacity and uncertain willingness of the governments to perform, as well as to the commitment of the private companies involved. Nevertheless, the Bank is willing to stake its reputational capital on this deal. It believes that the five years until the oil revenue comes on stream provide an opportunity to strengthen the institutions needed for the job.

Overall Outcome Assessment

Several questions must now be faced in light of the preceding discussions. What has the Bank helped Cameroon achieve in the context of forest resource management? How effective have Bank interventions been? How sustainable are the achievements? To what extent has the Bank implemented the 1991 Forest Strategy in Cameroon?

Achievements

Direct interventions of the Bank in the forest sector in Cameroon reflect the strategic shift that the Bank made in the late 1980s, when it moved away from investment operations in agriculture to policy-based lending. Thus, it attempted to reform the forest sector with dialogue supported by structural adjustment. The focus of the third Structural Adjustment Credit remains on: (1) improving the criteria for the allocation of timber harvesting rights (long-term concessions and annual cutting rights); (2) definition of a strategy for the apportionment of the areas to be awarded as concessions in a manner consistent with sustainable forest land use; and (3) reforming the forest taxation to improve the assessment and recovery of stumpage fees and royalties.

In justifying to the Board the release of the second tranche on June 24, 1999 (IDA/SECM99-396), the president noted the following achievements:

- *Harvesting rights.* The government adopted regulations containing a new set of criteria for the award of concessions and annual logging permits, and providing for the validation of the bidding process by an independent observer (these are codified in the Ministry of Forest's *Arrêtés* No. 757 and 758 dated June 16, 1999).

- *Planning strategy for the allocation of new concessions and logging rights.* The strategy has been defined and approved by the Ministry of Forest (*Décision* No. 787 dated June 15, 1999). This will be monitored as part of the forest sector floating tranche.

- *Forest revenue enhancement.* A decree (No. 99/370/PM), signed by the prime minister on March 19, 1999, establishes a program aimed at improving the assessment and the collection of the stumpage fee and the royalty. There has been a significant increase in tax revenues.

- *Taxation and industrialization.* The government is aware that a complete log export ban may not be in the best interest of the country. Thus, it has decided to commission a technical independent audit of the sector to assess its economic and tax potential and submit to the National Assembly amendments to the 1994 law on the basis of the recommendation of the audit.

Furthermore, the Bank has succeeded in elevating the debate in Cameroon (and indeed in the Congo Basin) on the issue of sustainable management of forest resources. The Bank helped the Cameroonian authorities organize an international summit on the promotion of rational forest use in the Congo Basin.

How significant are these achievements? All of them were accomplished by the stroke of a pen. In fact, the Bank acknowledges that the thrust of the ongoing dialogue and the adjustment program is on narrowing the gap between the stated policies and implementation, as well as improving transparency and accountability in the sector. This raises the issue of the effectiveness of the chosen instrument to induce forest policy reforms in Cameron.

Effectiveness of the Structural Adjustment Program

To understand the limited effectiveness of World Bank interventions in the forest sector, it is useful to review the politics of the reform process. This review will reveal some of the strategic mistakes that the Bank may have made.

Ekoko (1997, 1998) provides an excellent account of the politics of the Forest Law reform in Cameroon.[32] He argues that the process led to a limited outcome for four reasons. First, the stakeholders in the forest sector and the policymakers failed to agree on a common vision. Second, key actors in the reform process had divergent interests that led them to take divergent positions. Third, partners such as the World Bank misunderstood the underlying dynamics of political and socioeconomic change in Cameroon, in particular the intricacies between power and forest. Fourth, the government of Cameroon lacked genuine commitment and the capacity to carry out the reforms.

One may distinguish at least eight main groups of stakeholders concerned with forests in Cameroon: multilateral institutions such as the World Bank and the IMF; bilateral donors (the main ones include France, Canada, and the United States); international NGOs; the government, including both the executive and legislative branches; the bureaucracy; logging companies (foreign and local); local NGOs; the academic community; and the ordinary people, including local communities, farmers, and forest dwellers. These groups are far from being homogenous. The Country Team explains that within the donor community, the brunt of the dialogue was carried by the Bank and the IMF. Large logging companies did not see eye to eye with the medium-size or small ones. Different branches of government or of the bureaucracy were at war with each other.

The making of the 1994 Forest Law involved mainly the government of Cameroon, the World Bank, logging companies, and some local and foreign politicians. The outcomes at the drafting stage, the debate at the National Assembly, and implementation are better understood by considering the conflicting interests of these actors and the distribution of decision power among them.

The draft law submitted for debate at the National Assembly came out of an interaction between the Bank (backed by international NGOs) and the Executive Branch of government, with technical assistance provided by a Canadian consulting firm in the context of Canadian institution building project.[33] There was occasional participation by some logging companies and the Canadian International Development Agency (CIDA). Some analysts have suggested that the World Bank dominated the drafting process (Ekoko 1998). The main objective of the World Bank was to correct the deficiencies of the previous law and to ensure that local community interests and the principle of sustainable forest management are taken into account.

The weak bargaining position of the Executive Branch of the government at this stage reflected the financial and political difficulties it was facing at the time. By December 1993, and as a consequence of failing to seriously confront the economic crisis that started in 1986, real GDP declined by about 30 percent and real per capita income dropped by 50 percent (1996 ICR for Credit 2627-CM: 2). The government faced severe liquidity constraints, as it was unable to meet its financial obligations both internally and externally. Salaries went unpaid for months and the government resorted to arrears accumulation with respect to domestic and foreign debt.

In addition, the country went through serious civil unrest between 1990 and 1992. The regime felt sufficiently threatened by this to impose military rule over most of the territory. The outcome of the 1992 elections, which the ruling party won, was also contested not only by the opposition locally but by some bilateral partners. For a while, the United States and several European governments, excluding France, marked their displeasure by trying to isolate the government.

The above financial and political difficulties make it clear that the Executive Branch of government was under duress when negotiating the content of the 1994 Forest Law. The views of the Bank prevailed at the drafting stage with respect to issues of concessions, community forestry, sustainable management, and reserves and parks. The National Assembly did not, however, see itself under the same duress as the Executive Branch and set out to undo, during the debate, aspects of the law not to its liking. Indeed, during the debate at the National Assembly new players emerged and the Bank lost its leverage (Ekoko 1998). That debate centered around two issues: the efficiency and equity of public auction for the sale of standing timber, and the log export ban. On both issues, the Bank has not yet secured a completely acceptable outcome.

Foreign firms dominating the private sector[34] kept a low profile during the formulation of the policy, but made sure to join the fight in earnest at the beginning of the implementation phase. They carried a "big stick" in the form of repeated threats to close up shop. In general, the private sector fought every aspect of the reform that they perceived as eroding the protection they have enjoyed for at least 50 years.

Even the government institutions in charge of the implementation of the new forest policy were in conflict. In particular, the Ministry of the Environment and Forests wanted the creation of a Forest Fund to finance sustainable forest management activities out of the proceeds of forest fees and taxes. The Ministry of Finance (backed by the IMF) opposed the idea. The fund was eventually created in 1996 but had to wait until 1998 for the determination of a source of funding.

An unfortunate outcome of this process is that ordinary people whose livelihoods directly depend on forest resources were left out of the decisionmaking. This led to an inequitable forest law. Even though an information campaign was organized, it was restricted to some localities with no amplification at the national level. In addition, the opinions of the locals who attended the meetings were not particularly sought. Thus, the communication strategy adopted for the reform process ensured the capture of the reform by some key actors.

The World Bank apparently committed strategic mistakes. The Bank seems to have neglected the fact that policymaking is fundamentally a political process at all stages, including formulation, legislation, implementation (including the choice or creation of administrative agencies and their operation), and even evaluation.[35] If the Bank knew this, it certainly did not devise a strategy to see its proposal through. Our evaluation of the strategic framework imbedded in CAS confirms this point. Instead, the Bank generally adopted a "command-and-control" approach vis-à-vis the Executive Branch of government as if this branch of government were omnipotent and, therefore, able to deliver everything it promised at the bargaining table. As noted above, the government was weakened by political and financial difficulties. The Bank should have anticipated this and taken steps to ensure that the legislative phase would produce desirable results.

The objectives of the Bank in trying to promote the interest of local communities were good, but it did nearly nothing to gather their views and design mechanisms that would ensure that those views were taken into consideration. In fact, this may erode the credibility of the Bank with respect to its overarching objective of poverty reduction. Given the importance of this objective and that a key determinant of poverty is inequality in the distribution of social decision power, the Bank should have insisted on this issue.

Another strategic mistake that the Bank may have committed in this context is not to have applied a forward-looking approach to policy and project design. This generally means that the policymaker looks ahead and considers the type of institutional environment that will constrain implementation and feeds this information into policy design. The Bank rightly recognized institutional weaknesses in Cameroon, but preferred to rely on technical assistance to deal with the issue. Failure to develop local institutions undermines the sustainability of any achievement in this context.

Extent of the Implementation of the 1991 Forest Strategy in Cameroon

A major shift in Bank strategy was the commitment to taking a multisectoral approach with stronger linkages to country economic and sector interventions. This approach implies that the Bank designs non-forest-sector policies and operations, taking into account their impact on forests. The Bank partially followed a multisectoral approach to the extent that it raised forest sector issues within the dialogue and attempted forest sector reforms on the basis of adjustment.

One of the most controversial positions of the new strategy was the ban on lending for commercial logging in primary tropical moist forests. Even though the Bank did not lend for commercial logging in Cameroon, the ban does not seem to have helped.

Four principal areas of direct interventions were outlined in the reformulation of the Bank's forest strategy: international cooperation, policy reform and institutional strengthening, resource expansion and intensification, and preservation of intact forest areas.

International Cooperation

- Support for concessional resources to assist projects that protect globally important biological diversity
- Exploration of the feasibility of using global transfers to protect forests for their sequestered carbon stocks
- Through the Consultative Group on International Agricultural Research (CGIAR), assist in the expansion of natural resource management concerns and forest research into the work of the International Agricultural Research Centers.

Box 2.2. Incredible Conditionality and Government Commitment

In general, the reasons for the limited success of adjustment are to be found in the design and implementation of such programs. A typical adjustment program is in fact a two-stage contract. The first stage of this contract involves two parties, the financier(s) (MDBs) acting as the *principal*, and the central government of the country acting as *agent*. In this contract, a loan is exchanged for a promise of policy reforms. Government commitment is essential. Indeed, *government commitment has emerged as a key determinant of development effectiveness*. In a strategic interaction, commitment is a supporting or collateral action taken to make a strategic move credible. [a]

The current structure of conditionality is inadequate to induce commitment and therefore lacks credibility (Collier 1997; Collier et al. 1997). Initially, conditionality operated on an "all-or-nothing" basis. Sixty or more reforms would be packaged into a single reform contract subject to the same conditionality, stating that failure to implement any one of the reforms on time would result in program suspension. Such an outcome would place the country in default on its debt service and trigger a crisis that could reflect badly on all involved.

Over time, it became apparent that this type of punishment would not fit minor breaches. In the absence of lesser penalties, financiers resorted to wavers (Cameroon got a few along the way). The governments then knew that they could always count on such wavers in case of noncompliance. This time inconsistency (facing the governments with different incentives before and after the loan receipt) thus weakened the incentive structure facing the governments. The pressure to grant wavers may be political. It may also stem from the fact that some financiers get locked into defensive lending, having to lend more and encouraging others to lend to service their original loans.

A new style of lending has emerged in the mid-1990s to deal with the deficiencies of the big bundle approach. This approach is based on single tranche operations where reforms are priced individually and there is no payment in advance of implementation. There are two issues to consider in this approach. First, nothing can keep the same reform from being put back on the table year after year. Second, the approach reveals clearly that the government does not own the reforms. As Collier (1997:63) puts it: "*If donors price reform, they buy them and governments sell them. Who then owns the reforms?*"

The second stage of the contract involves the central government and the private sector/civil society. In a given society, the government does not act in a vacuum. Its actions call for a reaction from the private sector or civil society. Therefore, results on the ground depend on the nature of the strategic interaction among various socioeconomic agents subject to conflict and cooperation[b] within a governance structure.[c]

Collier argues that, if government commitment and good governance are so crucial for the success of policy reforms, then there is a need to reflect this in the design of conditionality. A basic approach could be to shift the basis of conditionality from promises to a retrospective assessment of sustained performance.

a. In a game, players actions are called moves. A strategic move is meant to alter the beliefs and actions of others in a direction favorable to oneself. Such a move purposefully limits the freedom of the mover. Credibility is a problem associated with all strategic moves. In this context, it is important to remember that a strategically aware opponent will expect the other party to mislead him and will therefore not be influenced by actions he perceives as being put on display for his benefit (Dixit and Nalebuff 1991: 120).
b. Strategic interaction means that choices by socioeconomic agents are interdependent.
c. In the abstract, a governance structure involves the provision of three elements: (1) universal and abstract rules; (2) enforcement institutions; and (3) mechanisms for resolving conflicts over the rules and their enforcement. Governance capacity is directly linked to the extent of the state's autonomy vis-à-vis the competing interests in society (Fristchtak 1994).

Policy Reform and Institutional Strengthening

- Assist in identifying and rectifying market and policy failures that encourage deforestation and inhibit sustainable land use.
- Assist with resource inventories and systems for continuous resource assessment.

Resource Expansion and Intensification

- Finance creation of additional forest resources and the expansion and intensification of management of areas suitable for sustainable production of forest products.
- Support family farm and farm group forestry including women's groups, especially on degraded forest lands.
- Establishment of plantations outside areas of intact forests.

Preservation of Intact Forest Areas

- Support for the expansion of forest areas designated as parks and reserves
- Support for the effective management and enforcement of parks
- Adoption in moist tropical areas of a precautionary policy toward forest utilization
- No financing of commercial logging in primary tropical moist forests
- Rigorous environmental assessment of infrastructure projects in tropical moist forests
- Implementation of strategy targeted to the 20 countries (accounting for 85 percent of tropical moist forests) whose forests are seriously threatened.

The Bank further indicated that lending operations would distinguish between projects that are environmentally protective or oriented toward small farmers and all other forest operations (such as commercial plantations). The first two types are to be considered on their own social, environmental, and economic merits. Other forest lending would depend on a solid commitment by government to sustainable and conservation-oriented forestry. If such a commitment were not evident, Bank support would be restricted to operations that help countries to achieve a sustainable forest sector. These operations would include, *inter alia*, the development of forest conservation and development plans and capacity for assessing social, economic, and environmental considerations in forest utilization.

Table 2.3 presents a summary evaluation of the implementation of the 1991 Forest Strategy in Cameroon. It is clear that the achievements fall short of expectations when judged against the overall framework

Table 2.3. Summary Evaluation of the Implementation of the 1991 Forest Strategy in Cameroon, 1991–99

Strategy Implementation

Did the Bank forest strategy for the country change after 1991?[a]	Yes
Was change attributable to the 1991 Forest Strategy?[a]	Yes
Was the Bank's post-1991 forest strategy for the country responsive to the needs articulated by the country?[a]	No
Consistency of Bank strategy	
Was the Bank strategy consistent with the CAS?[b]	Predominantly
Did the country have a forest policy consistent with the Bank's strategy?[a]	No
Did the Bank follow the principles of its involvement in the sector?[b]	
Multisectoral approach	Partly
International cooperation	Negligibly
Policy reform	Partly
Institutional reform	Negligibly
Preserving natural forests	Predominantly
Resource expansion and intensification	Negligibly
Were participatory approaches Implemented?[a]	Negligibly
Was the 1991 strategy implemented?[b]	Negligibly

Nature of Bank Interactions

The forest sector strategy was implemented through:[b]	
CAS	Partly
ESW	Negligibly
Policy dialogue	Partly
Lending to forest sector	Not applicable
Lending to forest-related sectors	Partly
Forest conditionality in adjustment lending	Mostly
Bank application of safeguards	Unclear

Bank Outcomes

Bank's forest sector strategy from country perspective:[c]	
Relevance	Modest
Efficacy	Negligible
Efficiency	Negligible
Is the impact of the Bank strategy in the country sustainable?[a]	Unclear

The Bank's Impact

Did the country improve its forest cover?[a]	No
Did the country improve the way it addresses forest sector issues?[b]	Negligibly
What degree of impact did the Bank strategy have on the poor?[c]	Negligible

Relevance for Future Strategy

Does the 1991 Forest Strategy seem relevant from the perspective of the country?[d]	Partly
Is there government demand for Bank involvement in the forest sector?[a]	No
Is there demand from NGOs, the private sector, and professionals for Bank involvement in the forest sector?[a]	Unclear
How was the country's forest policy embedded in its overall growth and poverty alleviation strategy?[e]	Poorly

a. Rating choices: Yes, No, Not Applicable, and Unclear.
b. Rating choices: Predominantly, Mostly, Partly, Negligibly, Not Applicable, and Unclear.
c. Rating choice: High, Substantial, Modest, Negligible, Adverse, Substantially Adverse, and Unclear.
d. Rating choices: Substantially, Partly, Negligibly, No, and Unclear.
e. Rating choices: Very Well, Well, Poorly, Very Poorly, Unclear.

laid down by the 1991 Bank strategy or even by the recommendations made in the 1989 agricultural sector review. Our assessment is that the extent to which the strategy was implemented is negligible.

With respect to biodiversity conservation, there is an ongoing (World Bank-implemented) GEF biodiversity project for US$12.39 million (including a GEF grant of US$5.96 million and contributions from France, the Netherlands, Germany, and the United Kingdom). The project was approved in March 1995 and became effective in December 1995. The two key objectives of the project are: (1) Help the government of Cameroon protect a significant amount of the country's biological diversity through a network of carefully managed national parks, reserves, and ungazetted sites; (2) Strengthen key national institutions concerned with research, planning, and coordination of biodiversity conservation activities at the national level. Two reviews of implementation progress, including the midterm review, found the outcome unsatisfactory due to poor design and lack of a national biodiversity conservation strategy and of adequate law enforcement as it relates to protected areas (favoring the expansion of logging and poaching).

New Developments

The Africa Region of the World Bank is currently preparing $US60 million Forest and Environment Sectoral Program to promote a coherent national programmatic approach to forest and environmental issues and the establishment of a stable, multi-donor, long-term support for the sector. The Project Concept Document identifies the following issues that must be addressed in order to close the gap between the laws and regulations on the books and reality on the ground:

The forest patrimony continues to degrade; forest regulations remain inapplicable or not applied; the relationship between the state, the private sector and civil society is confused and at times conflictive. Forest contributions to overall GDP remain below potential. At the heart of these difficulties are a weak political commitment to the 1994 law, and inadequate institutional capacity.

Thus, the key focus of the program is to strengthen both the public and private sector capacity to collaborate toward the goal of managing national forest and wildlife resources in a manner that is socially, economically, and ecologically sustainable. The program intends to push for institutional reforms with emphasis on field implementation of the current legal and regulatory framework. The program will also establish intersectoral linkages with *poverty* reduction and *governance*.

Four main components are planned:

- *Institutional Reform.* This component will focus on national land use planning and the improvement of forest control systems. The reform entails a redistribution of roles between the public and the private sectors with the goal of improving the efficiency of various forest and environment departments. The component will help the government implement and further refine the sectoral policy and institutional reforms initiated over the past 10 years. In order to improve sectoral governance and performance, the institutional reform component will foster the involvement of the private sector, and nongovernmental and community-based organizations.
- *Forest and Wildlife Resources Management.* This aspect of the program will comprise two subcomponents, one seeking to improve the management of production forests and the other to improve the management of savanna lands. In the first case, the program will help strengthen the capacity of all parties to undertake and oversee the implementation of forest management plans. In the savanna lands, the program will help build the capacity of local communities, create community-managed masterplans for fuel-wood exploitation, and develop the capacity of the private sector to engage in wildlife-related activities such as ecotourism and game ranching.
- *Biodiversity Conservation and Protected Area Management.* This component of the program seeks to strengthen government capacity to protect biodiversity in designated areas in collaboration with national and international partners, including the private sector, national and international NGOs, and the local population.
- *Participation in Global Environmental Initiatives.* The component is intended to help Cameroon meet the challenges and reap the benefits from active participation in international mechanisms designed to address issues related to biodiversity, desertification, and climate change at the global level.

In principle, it makes sense to build institutions where they are missing and strengthen capacity where it is weak. The key issues remain commitment and ownership. The designers of this program identified the Central African Forest Summit in Yaoundé as a hopeful sign. The design team also identified the following hopeful signs: (i) the Central African Forest Summit in Yaoundé on March 17, 1999; (ii) the joint initiative by the Ministries of Forests and of Finance to combat tax evasion in the forest sector; (iii) ongoing, massive staff transfer to inhibit corrupt practices; (iv) the purchase, out of the national budget, of equipment such as computers and motorcycles to facilitate enforcement of forest and tax rules and regulations; (v) the adoption and publication of a planning strategy to eliminate corrupt practices in three years; and (vi) the adoption of new bidding procedures in lieu of discretionary award of forest concessions.

An attractive feature of this program is that the design team seems determined to confront head-on the fundamental issues that have limited the success of Bank's interventions in the forest area. The Concept Document acknowledges that "*forest dialogue and policy implementation have been constrained by lack of consultative processes and the inability of the government and its development partners to establish participatory processes to inform concerned parties, and involved them in the decision and implementation process.*" To deal with this issue, the program intends to set up participatory mechanisms, such as roundtables, targeted communication campaigns, and consultation and feedback channels involving all stakeholders. The key challenge we see here is the design and implementation of mechanisms that entail a meaningful role for the local communities.

3

Toward a More Effective World Bank Strategy in Cameroon

Cameroon's forest resources, one of the country's greatest riches, have played and continue to play a significant role in its socioeconomic development. The forest resources of Cameroon are subject to four principal land use categories: large-scale industrial plantations, smallholder agriculture including community forests, production forests, and parks and reserves. The Bank—over the years, and to varying degree—has had an impact on all four of these land use categories through its policy dialogue and project lending. In the 1950s, 1960s, and 1970s conversion of moist forests to smallholder coffee and cocoa agroforests encouraged by Bank project lending resulted in relatively equitable economic growth averaging 3 to 4 percent. In recent years, timber exploitation has overtaken coffee and cocoa production as the most important economic activity in the moist forests. Cameroon is now considered the leading African exporter of tropical timbers, with over US$270 million in annual sales.

Until the late 1980s, the Bank's view was that forest resources were unused, or at best underused, and a recurring theme was that the long-term economic outlook "depends on making use of its vast natural resources particularly agricultural and forestry" (World Bank 1989). Policy recommendations were to remove the constraints to both greater expansion and intensification of these resources, which included transportation and port infrastructure, marketing arrangements, credit, and land tenure. By 1970, land tenure laws had been modified to permit the expropriation of lands in the public interest, providing a necessary institutional framework for nine Bank projects that converted forest lands to industrial plantations in the southwest of the country between 1967 and 1985. The Bank also supported and continues to support the transportation sector with projects that contributed to a reduction in marketing costs and increased Cameroon's competitiveness in world markets for both timber and export crops. However, even as early as the 1960s there was a concern for rational exploitation of these resources, with recommendations calling for forest inventories and upgrading of technical skills of forest sector staff to support improved management. Forest reserves for conservation were recommended for 20 percent of the moist forest area. From the standpoint of Bank strategy, the critical question was whether unused tropical forests would be turned into sustainable agricultural and forest production systems or "mined" into a state of degraded vegetation. This question is still pertinent and evident in Bank policy dialogue. A projected 80 percent increase in forest product exports by 2004 in the 1996 Country Assistance Strategy indicates that the continued development of Cameroon's forest resource is a major component in the country's and the Bank's strategy for economic growth. At the same time, much has been accomplished recently in formulating policies and developing a legal framework in an effort to ensure that these resources will be exploited in a sustainable fashion. Capacity building to implement and enforce these policies remains a formidable task.

In 1991, the Bank revised institution-wide strategy on tropical forests to address growing environmental concerns over the impact of Bank policy recommendations and project lending on the sustainable use of tropical forests. Among the major shifts in the policy was advocacy for a multisectoral approach to address the environmental issues of degradation and deforestation.

In Cameroon, smallholder agriculture is held to be the major source of deforestation, and therefore any proposed multisectoral approach for addressing deforestation must start with agriculture. The revised strategy, while according a dominant role in the deforestation process to agriculture across the tropics, actually had very little to suggest in terms of dealing with the problem.

We argue for a proactive policy-led effort to intensify perennial crop and food crop systems in order to deflect further advance of the forest margin. We also suggest that perennial crop agroforests are more sustainable agronomically and can provide some portion of the environmental services of tropical forests as compared to fallow-based food crop systems. At the same time, given the current state of underdeveloped rural food markets, food crop systems managed by women will remain an integral component of farming systems for the foreseeable future and therefore must be subject to policy-led intensification as well.

The necessary elements for such a strategy are grouped around the provision of improved price incentives and strengthening institutions. A viable and dynamic research and extension system capable of responding to farmer's demands and generating appropriate solutions is one of the main requirements. In the typical circumstances of the Congo Basin, where the farmer's scarcest resource is labor, technology systems typically economize on labor; therefore, to be viable, solutions must ensure that the farmer's return to labor is increased.

The differentiated labor roles of men and women in these systems and variation over the agricultural calendar have direct bearing on labor productivity and must be taken into consideration. Land and labor productivity are sometimes increased simultaneously by a given innovation. For instance, the fertilizer-seed systems of the green revolution increased both land and labor productivity and are credited with having deflected vast areas from agricultural conversion. Labor productivity measured in economic returns can be also increased through Pigouvian subsidies and taxes to correct for production externalities. Fertilizer subsidies in the Congo Basin take on a new perspective when one considers the potential economic costs in a carbon market of a small-holder burning her fallow field and releasing 100 to 135 tons of CO_2 annually into the atmosphere.

Market mechanisms such as ECO-OK labeling and the fair trade movement are attempting such corrections, albeit on a small scale and largely without the support of large donors like the World Bank. For areas that are already significantly degraded, institutional mechanisms for transferring carbon credits to smallholders are needed to provide new planting incentives for reforestation through agroforests. Supportive rural institutions such as credit markets, input markets, and the multiplication and distribution of improved planting material must also be strengthened.

Intensifying production in the forest zone is not a new idea in Cameroon. As early as the 1930s the French were searching for ways to intensify cocoa production in Cameroon. Most of these efforts did not achieve their goals, including the Bank-sponsored cocoa projects. What is different today, however, is the much-improved economic environment for intensification. Although the policy sequencing of reforms within the agricultural sector was less than ideal for producers and the environment, producers are now showing renewed interest in their perennial crop systems following the removal of implicit producer taxation administered by the now defunct marketing board. The infusion of new competition into export crop marketing has also reduced marketing margins and increased producer incentives. These incentives need to be augmented even further, and technologies and institutions developed, to exploit Cameroon's natural comparative advantage more efficiently. This will require capacity building in the public sector and the exploration of new institutional arrangements for the provision of services and inputs by the private sector, NGOs, and farmer organizations.

A promising development for intensification is the greatly increased contact between farmers and extension agents as a result of the World Bank extension project. In many ways the project is quite innovative— for instance, its focus on women farmers and targeting women in recruitment of extension agents. Explicit support for the burgeoning farmer movement in Cameroon was provided by the liberalization of export markets in the early 1990s, which established an improved legal framework for organization. The approach is also much more participatory than previous extension efforts. There is, however, concern that the supply of innovations available for farmers in the humid forest zone may not be sufficient because of the problems that have beset agricultural research over the last 15 years. Another crucial element in the sustainability of this project will be the capacity of government to maintain the structure after the project is finished, and in particular the mobility of 1,700 agents, which has been greatly increased by the project provision of motorcycles.

A multisectoral approach for addressing the deforestation and degradation problems is not evident in the current set of Bank interventions. The appraisal document for the extension project makes no explicit mention of the problem, nor does it clearly outline a strategy for intensification in land-surplus, labor-scarce rural economies. A similar degree of neglect is to be found in the transportation project. Transportation infrastructure supported by the Bank has in certain instances led to increased pressures on forest resources, while in others it has decreased pressures by increasing market access of farmers in areas outside the forest zone. The approach was also not evident in the ill-advised sequencing of policy reforms in the agricultural sector, which resulted in serious depreciation of farmers' tree stock capital and increased deforestation as resources were shifted into extensive slash-and-burn food crop systems.

The second major environmental concern surrounds the degradation of timber and non-timber forest resources associated with the rapid increase in logging activity over the last 15 to 20 years. The Bank, which has supported only one ineffective forest project in Cameroon in the 1980s, has had most of its impact on the forest sector through policy dialogue conducted with the "big stick" of structural adjustment lending in hand. Achieving sustainable forest management is in many ways similar to intensifying agriculture; there is a need for technical innovations and increases in the stock of knowledge, as well as an appropriate incentive structure for adopting these practices. To date the adoption of improved management plans within the forest sector has been resisted as the economic benefits of sustainable management practices are not widely perceived within the industry. Findings from IRAD, CIRAD, and Tropenbos on sustainable management are indicating that the process is also likely to be knowledge-intensive and will require significant training of staff charged with approving, supervising, and enforcing the management plans of logging concessions. Loggers will also need new information and technology to implement sustainable practices.

There is also concern over the limited forward and backward linkages in the sector and the high degree of foreign ownership and the implications for sustainable development. The political economy surrounding this set of issues was crucial in determining the outcomes associated with the Forestry Law.

The capacity for implementing and enforcing policy in both the agricultural and forest sectors is weak. The overall capacity of the public sector was significantly weakened by the fiscal compression that the government implemented as part of its structural adjustment program. The government, for political reasons, chose to reduce salary levels across the board instead of reducing staff numbers. Thus, instead of increasing efficiency of the public sector there was an overall decline in efficiency with the drop in wages. While the civil service has slowly been decreased in size through hiring freezes and early retirements, the problem of a low-paid, inefficient civil service remains a major obstacle to the strengthening of institutions required for the sustainable development of forest and agricultural resources. Without support for institutional development, the significant gains achieved in environmental and forest policies will remain only paper policies.

As the experience with the Structural Adjustment Program shows, it is useless to design a strategy if it cannot be implemented successfully on a sustainable basis. Feasibility and sustainability require political and non-political resources. The fact that the outcome hinges on the interaction among different interest groups raises a host of challenging issues relating to borrower ownership and capacity, donor coordination, and the appropriateness of the adjustment instrument for forest policy reform.

It is very clear that the use of conditionality could not induce the commitment of the government of Cameroon for the implementation of forest policy reforms. This is understandable given the weakness of incentives facing the government. Contributing factors to this weakness of incentive include the lack of donor coordination, the use of waviers, and the fact that conditionality is based either on a promise or on the delivery of "stroke of the pen" measures. True commitment requires costly actions that are hard to reverse. This suggests a shift in the design of conditionality. Instead of basing tranching on policy content, it should be based on an acceptable policy process, one that involves partnerships outside the central government.

Some observers explain this weakness of incentives by pointing out that government's behavior in the forest sector is driven mostly by incentives in the form of the vast private gains and political patronage that can be secured in the context of uncontrolled, unsustainable, and, at times, illegal forest exploitation. Such a situation presents a huge challenge for the Bank to offer countervailing incentives.

Such partnerships can be built through participatory economic and sector work (involving local communities, academia, and NGOs) and proper information dissemination (involving local mass media and communities). Collaboration in economic and sector work may improve communication between the Bank and the country and establish a sense of partnership with key stakeholders. This setting would allow the Bank to gain better knowledge of and sensitivity to the client's circumstances. This approach would make Bank work more transparent and possibly enhance its credibility (as an honest broker) among stakeholders. It is likely that, when work is done in partnership, political and social issues can be raised, confronted, and dealt with up front. In the end, one has to face tradeoffs between the best assessment of experts and the sociopolitical interests operating in the relevant sector. It may be more important to get the process right than to have a best solution that cannot be implemented.

New Strategic Direction

There are indications that the Bank is now moving in this direction: witness the emphasis placed on institutional development in the conception of the new lending program in the sector. In addition, the Africa Region of the World Bank is currently drafting a Joint Operational Strategy for Forests and Biodiversity. The strategy stakes the relevance of the Bank's interventions in the continent on the role it plays in the sustainable use and conservation of Africa's natural capital. This stems from the premise that Africa's destiny is linked closely to the effective use of and conservation of its forests, water, agricultural land, and biodiversity.

The Joint Operational Strategy is designed to be consistent with and to reinforce the following key priorities: (1) improving governance and capacity building; (ii) investing in people through health and education; (iii) increasing competitiveness and diversification of economies; and (iv) reducing aid dependence and debt, while strengthening partnerships.

A successful implementation of the above strategy would mark, indeed, a new era in the development history of the African continent. *Affaire à suivre.*

Annexes

A. The 1991 Forest Strategy

The World Bank Forest Strategy sought to address rapid deforestation, especially of tropical moist forests, and inadequate planting of new trees to meet the rapidly growing demand for wood products. These twin challenges were the consequence of five forces:

- Externalities that interfered with the free interplay of market forces with the potential to bring about socially desired outcomes
- Strong incentives to cut trees
- Weak property rights in many forests and wooded areas
- High private discount rates for those encroaching on the forests and
- Inappropriate government policies, particularly concession arrangements.

The Bank's strategy, therefore, promised to promote the conservation of natural forests and the sustainable development of managed forest resources. The strategy it outlined consisted of policies to alleviate poverty, improve forest zoning and regulation, correct private incentives, and increase public investments. The strategy also proposed reducing demand through investments in research and technology, increasing the supply of essentials through farm forestry, and increasing market efficiency. Government policies and programs, the strategy said, should aim to change the incentives and institutional structures that lead to excessive deforestation and inadequate tree planting and prevent the use of good practices in forest management. Under the strategy, international cooperation and

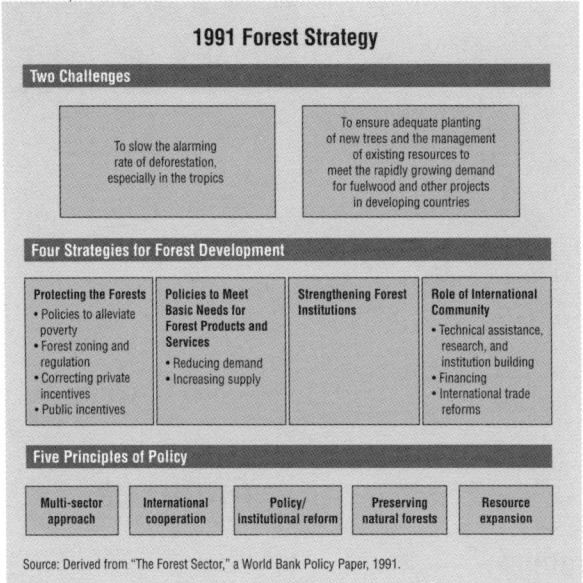

Source: Derived from "The Forest Sector," a World Bank Policy Paper, 1991.

assistance were to ensure that global externalities were internalized locally and that the efforts of governments and international organizations were to be coordinated.

Five principles were elucidated to underpin Bank involvement in the forest sector:

- Adopt a *multi-sectoral approach* in the design and implementation of forest operations.
- Support *international cooperation* in the formulation and adoption of legal instruments conducive to sustainable forest development and conservation.
- Promote *policy reform and institutional strengthening* by helping governments identify and rectify market and policy failures that encourage deforestation and unsustainable land use.
- Finance operations that lead to socially, environmentally, and economically sustainable *resource expansion and intensification.*
- Support initiatives that *preserve intact forest areas.*

Fulfilling this commitment required five things of Bank-financed projects:

- Adoption of policies and an institutional framework consistent with sustainability and a participatory approach to the management of natural forests
- Adoption of comprehensive and environmentally sound conservation and development plans based on a clear definition of the roles and the rights of the key stakeholders, including local people
- Basing commercial use of forests on adequate social, environmental, and economic assessments
- Making adequate provisions to maintain biodiversity and safeguard the interests of forest dwellers, particularly indigenous peoples
- Establishing adequate enforcement mechanisms.

B. Biodiversity Resources in the Forests of Cameroon

Biodiversity in certain of the moist tropical forest ecosystems of Cameroon is among the most extensive and unique to be found, both in Africa and across the globe. As such, these ecosystems are high on the protection and conservation list of international environmental NGOs.

Biodiversity can be defined as the variability within and among species. It has been produced by evolutionary processes over the course of millions of years. This biological stock of natural capital is the source of numerous food products, pharmaceuticals, and domestic and industrial products. Okigbo (1994) views the goal of biodiversity conservation as the management of resources to ensure that human needs are satisfied, while at the same time maintaining the biodiversity necessary for meeting the needs of future generations. Under this definition, the conservation of biodiversity entails both management for current use and preservation for the future.

Endemic Species

Cameroon's forests, ranging from incredibly wet evergreen forest in the rain shadow of Mount Cameroon to the semi-arid Guinea Savanna Woodlands of northern Cameroon, exhibit most of the variation in vegetation types found in Africa. Cameroon is also one of the few places in the world where tropical montane forest systems are found. These are particularly important centers of plant and faunal endemism. On Mount Cameroon alone, over 45 endemic plant species have been described (IUCN 1994). Cameroon is also an important site for the so-called charismatic fauna such as elephants, western lowland gorillas, leopards, lions, forest buffalo, bongo antelope, and chimpanzees, and has the only remaining population of black rhinos in West Africa. Elephant, gorilla, leopard, and chimpanzee populations are largest in the relatively intact forests of the southeast. The biodiversity richness of Cameroon's forests is exceptional among African countries. Unlike much of the rest of Africa, parts of the Cameroonian rainforest remained intact during the cool, dry weather of the Pleistocene epoch, which explains their relative richness (IUCN 1994).

According to Okigbo (1994), the estimated 10,000-plus plant species in Cameroon are only exceeded in number by the much larger Democratic Republic of the Congo (ex-Zaire) in West and Central Africa. Satabié (1995) estimates endemic species at over 160 within the most important plant families: Orchidaceae, 30; Podestemaceae, 18; Lauraceae, 17; Melastomataceae, 14; and Cesalpiniaceae, 11. Members of the fungi, lichen, fern, gymnosperm, and angiosperm plant classes are found. Fungi and lichens are estimated at 277 species, of which 257 have been described botanically in the "Flore du Cameroun." Less represented are the gymnosperms, vegetatively primitive plants that include the economically important and intensively harvested wild forest species *Gnetum africanum* and *Gnetum bulcholzianum*. As for the vegetatively evolved angiosperms, they are estimated at between 8,000 and 10,000 species, of which only 2,150 have been botanically described in detail.

Differences in Biodiversity Across Ecological Regions

The National Environmental Management Plan (NEMP) divides the country into four major ecological regions—the Sudano-Sahelian zone in the far north, the savanna zone, the coastal-marine zone, and the tropical forest zone. Within these regions, NEMP further subdivides into 10 ecological zones, dividing the tropical forest zone into the degraded forests of the Center and Littoral Provinces and the dense humid forests of the southwest, south, and east. Letouzey (1986) provides a more disaggregated classification of the humid forest, dividing the zone into 16 evergreen types and 4 semi-deciduous types. An intermediate classification building on Letouzey's divisions groups the humid forest zone into:

- The dense moist evergreen Atlantic forest, which is further subdivided into the coastal and Biafran forests
- The Cameroon-Congo forest
- The semi-deciduous forest
- Mangrove forests (Gartlan 1989).

For the purposes of this study we choose to use this latter classification because of important distinctions in biodiversity richness across these classifications (table B.1).

The coastal forests lie between the mangrove forests and the Biafran forests and are characterized by an abundance of the dense hardwood *Lophira alata* (Azobe), which is one of the most important commercial timber species. The abundance of this species is probably due to human influence as it thrives in clearings. These coastal forests, because of their access to the Douala port, have been logged over several times and are now a relatively depleted source of timber resources.

In terms of floral diversity the most important of the humid forest ecotypes are the Biafran forests. Over 200 plant species have been counted within a 1,000 m² transect of the Biafran forest. This represents a higher plant diversity than any other forest in Africa or Southeast Asia and is

Table B.1. Floristic Characteristics and Extent of Moist Forest Ecotypes in Cameroon

Type of forest ecosystem	Area (km^2)	Biological characteristics
Afromontane	725	Height 15–25 m, evergreen, high endemism but relatively low species diversity, important reserve of Prunus africanus.
Guineo-Congolian	267,000	
Submontane forest	3,775	Lies between 800–2200 m in elevation, increasing diversity of epiphytic flora with elevation, Prunus africanus found at higher elevations, biology of ecosystem not well known compared to lowland and montane forest systems.
Dense humid evergreen Atlantic forest (Biafran)	54,000	Very high floristic diversity with marked endemism, flora with affinities to South American forests, center of diversity for various plant taxa including the genera *Cola, Diospyros, Garcinia,* and *Dorstenia.* Biafran forest type characterized by gregarious associations of *Caesalpiniaceae.*
Dense humid Cameroon-Congolese forest	81,000	Intermediate in floristic diversity between the Atlantic forest and the semi-deciduous forest, flora affinities with Congo Basin forests. Important ecosystem for large primates and elephants.
Dense humid semi-decidous forest	40,000	Often fragmented, subject to fire during the dry season, particularly rich in commercial timber species, although less biologically diverse than other tropical forest types.
Mangrove	2,434	Dominated by red mangrove *Rhizophora racemosa* with lesser occurrence of white mangrove *Avicennia nitida,* found on the coast to the east and west of Mount Cameroon, some ecological damage (eutrophication) as a result of large-scale industrial plantation fertilizers and pesticides. Tree growth is correlated with rainfall. In coastal areas to the north of Douala, red mangrove heights can reach 25 m.

Source: IUCN 1994.

greater than most South American forests (Gartlan 1989). This forest type is also a center of genetic diversity for important genera such as *Cola*, *Diospyros* (ebony), and *Garcinia* (which includes the bitter cola). These forests were Pleistocene refuges and are therefore also characterized by a high number of endemic species. There are also certain plant species that show affinities to the forest communities found along the Atlantic coast of South America. These forests are under high human population pressures.

The Cameroon-Congo forest has a much lower rate of plant endemism than the Biafran forests but is important in term of biodiversity because of relatively low population pressures. Most of Cameroon's elephants reside in these forests. Densities of 2.8 animals/km^2 and 3.2 animals/km^2 were estimated for elephants and lowland gorillas within the 225,000-ha proposed park of Boumba-Bek, resulting in estimated park populations of 6,524 ± 2,586 elephants and 7,233 ± 2,097 gorillas (Ekobo 1995). These are very high densities and represent an important ecotourism potential that has yet to be exploited.

The amount of diversity with regard to fauna and fish species, as with plant species, is general higher in the humid forest zone of Cameroon than elsewhere. Of the 250 mammalian species in Cameroon, 162 exist in the humid forest (65 percent), with 132 of these species only found in this habitat. Similarly, of the 542 fish species identified, 294 exist in the freshwater resources of the humid forest zone, of which 78 are endemic to Cameroon. The number of fish species in Cameroon is greater by twofold than the total in all of Europe. Clearly the biodiversity reserves of the moist humid forests of Cameroon are of high intrinsic value.

Crop Genetic Diversity

Maintaining the genetic diversity within indigenous and landrace[1] varieties of agricultural crops is critical for food security and productivity growth in agriculture. This diversity provides the building blocks essential for increasing crop yields. The protection of within-species genetic variation is a key factor in the long-run survival of species and is a major concern in plant genetic resource programs for crop species. This genetic variation is contained within the germplasm of the plant species. The germplasm of all species manifests differing degrees of variability, expressed as traits or characteristics that are the result of genetic mutations. It is this variability (the culmination of thousands of years of genetic mutations) that supplies plant breeders with the building blocks necessary for improving agricultural crops. As environmental change has accelerated along with population growth and rural development, new plant diseases and pests, declining soil fertility, and global climate change have also evolved. So far modern agricultural science has successfully combated these changes, in large part by drawing on the stock of genetic resources to develop new, improved plant varieties and animal breeds. The ability of science to continue to address rapidly evolving changes in agricultural systems through genetic improvement requires that the foundation of this genetic variability be maintained.

The Biafran forests of Cameroon are recognized as a center of origin for yam, *Dioscorea*, which is the among most important commercial food crops in the humid forest zone of West Africa. These forests are also the center of dispersion for the oil palm *Elaeis guineensis*, and its major pollinator *Elaeidobuis kamerunicus*. This weevil, when introduced to Southeast Asia in 1981, increased oil production by almost 20 percent (IUCN 1994). The Biafran forests are also centers of endemism for the ginger and arrowroot family (Zingiberaceae). Also native are several species of coffee, including robusta coffee *Coffea canephora*, grown by smallholders throughout West and Central Africa.

There are numerous ways in which biodiversity can be threatened by the activities of rural populations. Among the actions and processes impacting biodiversity in the humid forests of Cameroon are deforestation, overharvesting of non-timber forest products, population growth, structural change in agricultural systems, and cultural change.

In areas where population growth has been rapid and deforestation has occurred, wild and/or closely related species may be threatened (for instance, *Coffea canephora*). Deforestation is a particular threat in centers of plant origin, where genetic introgression with wild and weedy relatives is a major source of genetic variation. Crop biodiversity can also be lost as farming systems and social relations evolve in the development process. Two interesting examples come from the Lékié Division of the Forest Margins Benchmark (see Annex E). In this densely populated portion of the benchmark, it is very difficult to find areas where the *ngon* melon (*Cucumeropsis mannii*) is still produced; instead, its place in the farming system has been taken by the *omgbalag* melon (*Cucumis sativa*). Because of shortening fallow periods, *ngon* melon can no longer be grown; *omgbalag* melon, in contrast, can be grown in short-fallow fields.

The second example is the erosion of yam genetic resources as a result of cultural change. Yam has a low seed-to-yield ratio and is difficult to store; thus, when yam is not cultivated, it can rapidly disappear from farming systems. The decline in yam production among the Eton of the Lékié has been attributed to changes in household structure that occurred with the adoption of Christian beliefs (Guyer 1984). When Eton marital relations became monogamous (in line with a Christian moral system), household female labor supplies declined, and the labor-intensive production of yams was often no longer viable. While it is still possible to find yam fields in some parts of the Lékié, they are much less common than in the past and local landrace diversity undoubtedly has been lost.

Conserving the biodiversity of non-timber forest products, wildlife and fisheries is extremely important for sustaining livelihoods, particularly in remote areas that may be marginally productive agriculturally. Non-timber forest products such as wild fruits and foods, building materials, and medicinal plants are very important in the subsistence strategies of rural households, particularly in the areas where forests have remained relatively undisturbed. The degradation of these resources with increasing population pressure is evident as one moves toward the northern portion of the Forest Margins Benchmark, where population densities are highest. Households in the less densely populated southern portion of the Forest Margins Benchmark devote significantly more time to hunting, fishing, and gathering of wild foods and medicinal plants than households in the Yaoundé block (Gockowski and Baker 1996). This greater reliance is related to the underdevelopment of roads, medical facilities, agricultural extension, and the lower population pressures of the region.[2] Because of these factors, the economic pursuit of game, fish, and non-timber forest species is relatively more important in areas of low population pressures. A major question facing researchers and development agents is how to maintain a greater proportion of this biodiversity as the development process proceeds.

In addition to the value of household consumption, the revenues generated from these products can be a significant source of household income, especially for resource-poor households in areas were these resources are still relatively robust.[3] Among the most important markets for NTFPs are bush meat, bush mango (*Irvingia gabonensis*), njansan (*Ricinodendron heudelotii*), kola nut (*Cola acuminata*), bitter kola (*Garcinia kola*), essock (*Garcinia lucida*), African pear (*Dacroydes edulis*), *Gnetum africanum*, rattan (*Ancystrophyllum* spp.), and ndin (*Monodora myristica*) (Ndoye 1995a, 1995b).

The conservation of Cameroon's plant genetic resources and biodiversity is an important national concern, vital to sustainable rural development. Rural development everywhere has been accompanied by exploitation of plant genetic resources. In order to exploit these resources they must be guarded and conserved. At the same time, there is great international interest in the conservation of Cameroon's unique biodiversity. The immense diversity of Cameroon, the unique and rare nature of certain ecological sites, and the animal and plant species endemic to these areas are global goods that require support not only from the government of Cameroon, but from the international community as well.

C. Estimates of Forest Cover and Deforestation

Estimates of forest cover in Cameroon depend somewhat on the classification scheme and methodology.[4] Many are based on extrapolations from distant baseline estimates using assumptions about uncertain rates of deforestation. Forest cover of all types in Cameroon in 1995 is estimated by the FAO (1997) at 195,980 km², or 42 percent of total land area (table C.1). Moist tropical forests with closed canopies are estimated by White (1983) to have once covered 376,900 km², which is more than twice Letouzey's (1985) estimate of 155,330 km² in 1985 and the FAO's estimate of 179,200 km² in 1980.

Millington and Pye (1994) used 1986 AVHRR remote-sensing analysis to estimate the extent of a wide range of land cover classes, including forest and non-forest, not just for Cameroon but for all of Sub-Saharan Africa (table C.2). Their classification provides evidence on the extent of deforestation. The five forest types, in descending order of spatial extent, are the mesophilous humid tropical forest (i.e., semi-deciduous forest), the ombrophilous humid tropical forest (Guineo-Congolian evergreen forest), tropical swamp forest, mangrove forest, and montane forest. The total area cover by these forests was 205,971 km², which together accounts for 44 percent of total land area. The extent of deforestation due to human influence is deduced from the category "cultivation and forest mosaic." Some 9 percent of land area is covered by this land class. Millington and Pye (1994: 10) note that "the key to understanding this extensive mosaic is that in most locations it has been greatly affected by human interference . . . It occurs in conjunction with the forest classes . . . and covers many ecological zones." As a result, "the tree species are quite varied. In Cameroon, species from the Atlantic Evergreen Forest

Table C.1. Estimates of the Extent of Forest Cover in Cameroon

Source	Area	Year of estimate	Remarks
FAO (1997)	195,980	1995	FAO definition of forest (see footnote)
White (1983)	378,900		Primordial
Letouzey (1985)	155,330		
Millington and Pye (1994)	205,971	1986	Remotely sensed AVHRR analysis
Singh (1993)	242,000	1970	
Singh (1993)	165,000	1990	
Smith et al. (1991)	202,650	?	(As reported in Dixon et al.)
Gartlan (1989)	200,000	1986–87	Moist, dense, closed-canopy forest
Gartlan (1989)	210,000	1980–81	

Table C.2. Land Cover Classes in Cameroon, 1986 AVHRR Data

Land cover class	Area km²	%	Growing stock tons (thous.)	%	Sustainable yield tons/yr (thous.)	%
Desert	105	0	0	0	0	0
Grassland	2,687	1	610	0	27	0
Semi-desert wooded grassland	105	0	35	0	1	0
Acacia wooded grassland	2,635	1	869	0	26	0
Plateau wooded grassland	790	0	261	0	8	0
Wooded grassland	3,530	1	1,165	0	35	0
Dry bushland and thicket	158	0	219	0	3	0
Moist bushland and thicket	632	0	1,069	0	13	0
Sahel-Sudanian wooded bushland	10,961	2	1,545	0	384	0
Bushland and thicket	11,751	3	2,834	0	400	0
Low woody biomass mosaic	685	0	1,472	0	43	0
Dry Sudanian woodland	34,779	8	92,860	3	2,677	1
Sudanian woodland and thicket	685	0	1,829	0	30	0
Sudanian woodland	21,974	5	58,671	2	1,011	0
Moist Sudanian woodland	65,289	14	174,322	6	3,134	1
Woodland	122,727	27	327,681	11	6,853	2
Cultivation and forest mosaic	42,525	9	71,612	2	808	0
Guinean woodland	59,915	13	100,897	3	1,138	0
High woody biomass mosaic	102,440	23	172,509	6	1,946	1
Mangrove	8,906	2	219,800	7	26,263	8
Montane forest	53	0	525	0	21	0
Mesophilous humid tropical forest	88,370	19	1,129,192	36	130,788	40
Humid tropical swamp forest	45,792	10	585,130	19	67,662	21
Ombrophilous humid tropical forest	62,750	14	658,185	21	92,870	28
Forest	205,871	46	2,592,832	84	317,715	97
Lakes	2,161	0	0	0	0	0
Cloud cover	14,649	3				
Total	466,606	100	3,099,103	100	327,019	100

Source: Millington and Pye 1994.

dominate (in the southwest), notably Caesalpinaceae," while in the Center Province species are from the Mesophilous Humid Tropical Forest. They go on to note, "The remaining forest in this class is inevitably secondary, with light-demanding tree species such as *Chlorophora excelsa*, *Khaya anthotheca*, *Musanga cecropioides*, and *Terminalia superba*. Cultivation throughout this class includes large areas of manioc. Other crops are grown in areas of secondary Annex forest." Among the areas of the forest zone characterized by this mosaic are the Lékié division (which is covered by the IITA ecoregional benchmark discussed below in the section dealing with agriculture), the Moungo division in the Littoral Province, and the Meme and Fako Divisions in the Southwest Province. Outside of the forest classes and cultivation and forest mosaic we run into ecological classes that, although altered by human influences, were never likely to have been moist tropical forests. Thus, if Millington's classifications are correct, there were 248,396 km^2 of moist tropical forest in the recent past, of which 17 percent no longer exists.

As with the estimates of forest cover, estimates of deforestation vary considerably. This is partly due to differences in definitions, and partly to difficulties in assessment. Rates of deforestation, of course depend on the choice of an arbitrary starting point. For example, if we assume that the process of deforestation began once the semi-nomadic nature of western Bantu society in southern Cameroon ceased with the German policy of sedentarization beginning around 1900, then the average annual rate of deforestation, using the estimate of Millington and Pye, would be 0.2 percent. If we assumed that at the start of this period there was instead 378,900 km^2 of forested area, as estimated by White (1984), then the rate of deforestation over this same period amounts to 0.9 percent annually. In another example, FAO (1988) gave a figure of 800 km^2 lost annually from 1981 to 1985, and then estimated a loss of 1,290 km^2 from 1991–95 (FAO 1997), while IIED (1987) estimated an annual loss of 1,500 km^2 and Singh (1993) estimated 1,370 km^2 from 1980 to 1990. In a 1986 study conducted in the preparation of the Tropical Forest Action Plan, a figure of 2,000 km^2 was given, half of which occurred in the dense moist forest and half in the woodland savannas (*Afrique Agriculture* 1990). Gartlan (1989) estimates a loss in dense moist forest cover of between 800 and 1,000 km^2. If we assume a base moist forest cover area of 200,000 km^2, these estimates result in annual deforestation rates ranging from 0.4 percent to 1.0 percent annually. Other rates reported in the literature also fall with in this range.[5] By comparison, in Indonesia and Brazil for 1990 to 1995, the estimated average annual forest loss was 10,840 km^2 per year (1.0 percent) and 125,540 km^2 per year (0.5 percent) respectively, while in West moist Africa and in Central Africa it was 4,920 km^2 (1.0 percent) and 12,000 km^2 (0.6 percent) respectively (FAO 1997).

In an interesting approach, Gaston et al. (1998) combine ecological data on carbon pool estimates for forest types with FAO forest cover maps from 1980 and 1990 and population density data to estimate the change in total carbon pools due to deforestation and degradation. In their study, degradation is defined as long-term human use that has a significant effect on biomass density of tropical forests. These uses include fuelwood gathering, logging, grazing, shifting cultivation, and anthropogenic burning. To estimate these effects they used forest inventories to calculate biomass reduction factors, which were then regressed on their corresponding population densities. They estimate that the aboveground carbon pool in Cameroon forests (FAO definition) declined by 416 Tg, which is equal to an average annual decline of 1.7 percent. Seventy percent of the decline is due to "degradation" and the rest to deforestation. This is a much higher estimate than other authors attribute to the sector.

D. The Yaoundé Environmental Summit of Central African Countries

Maintaining biodiversity in forest ecosystems is the focus of numerous environmental NGOs, both local and international, in Cameroon. The environmental movement has been successful in advancing the conservation agenda and integrating their concerns into national and regional policy dialogue. On March 17, 1999, the Yaoundé summit of Central African environmental ministers and heads of state was initiated and hosted by President Paul Biya of Cameroon and chaired by HRH The Duke of Edinburgh (the World Wide Fund for Nature's International President Emeritus). The convening of the summit exhibits the influence of the environmental movement as well as the sincere concern of the Cameroon government with maintaining the sustainable use of this resource in Cameroon. The main objective of the summit was to discuss the creation of new cross-border forest protected areas and to ensure that sustainable forest management and independent timber certification become a commercial and conservation reality. The summit witnessed the signing of the Yaoundé Declaration, which outlines the steps to be taken in order to ensure the integrity of forest ecosystems in Central Africa.

Prior to the conference, the World Wide Fund for Nature (WWF) circulated the steps that it hoped would be undertaken. These included:

- Establish a new transborder conservation initiative among Gabon, Cameroon, and Congo-Brazzaville.
- Enforce the existing tri-national network of protected areas in Cameroon, the Central African Republic, and Congo-Brazzaville.
- Donate two new Gifts to the Earth in Cameroon through the creation of two forest reserves in Boumba Bek and Nki in southeast Cameroon.
- Establish by presidential decree a trust fund in Cameroon to finance the effective management of forest protected areas.
- Adopt a national elephant management plan, including the protection of forest habitats, for Cameroon's elephants, most of which live in Cameroon's southeast forests.
- Work toward the establishment of regional certification standards under the auspices of the Forest Stewardship Council[6] to encourage sustainable forest management focusing specifically on Cameroon, the Central African Republic, and Gabon.

At the end of the summit, the Yaoundé Declaration was read by Ferdinand Leopold Oyono, Cameroon Minister of Culture, and included the following points:

- Accelerate the process of creating transborder protected areas, while inviting neighboring countries to participate, and while reinforcing the sustainable management of existing protected areas.
- Support conservation, sustainable management, and forestry research through permanent fiscal measures applied to the forest sector.
- Adopt harmonized national forest policies and accelerate the putting in place of management provisions, notably internationally recognized timber certification programs agreed to by the Nation States of Central Africa and develop the human resources needed for their effective implementation.
- Reinforce mechanisms leading to the increased participation of rural populations in the planning and sustainable development of their ecosystems and reserve adequate land for their economic, social, and cultural development.
- Encourage the greater participation of private enterprise in the sustainable management and conservation of forest ecosystems.
- Take measures to reconcile actions addressing forest ecosystems with those of other sectors, such as transportation and agriculture.
- Put in place concerted measures to halt poaching and all other non-sustainable exploitation in close collaboration with local populations and private enterprise.
- Promote and accelerate the processing sector and develop financial mechanisms to support the private sector in view of maximizing added value and creating new and remunerative employment, all the while guarding forest resources in a context of sustainable utilization.
- Promote forums for exchanging information at the national and subregional levels, encourage networks of research institutes and forest development stakeholders, and reinforce the coordination and cooperation among all national and international organizations implicated in actions and reflections concerning the utilization and conservation of biological resources and forest ecosystems.
- Put in place sustainable mechanisms for financing the development of the forest sector from sectoral revenues.

- Organize regular summits devoted to the conservation and management of forest ecosystems.

The Yaoundé Declaration was signed on March 17 by Paul Biya, President of the Republic of Cameroon; Denis Sassou Ngesso, President of the Republic of Congo; Omar Bongo, President of the Gabon Republic; Théodoro Obiang Nguema, President of the Republic of Equatorial Guinea; Ange Félix Patassé, President of the Central African Republic (CAR); Laurent Désiré Kabila, President of the Democratic Republic of Congo; and Idriss Deby, President of the Republic of Chad.

From the viewpoint of the conference organizers, the Yaoundé Declaration incorporates most of their points of concern, at least tangentially. The most important issue was the expression of support for establishing a network of transborder protected areas between southeast Cameroon, southwest CAR, northwest Congo, and northern Gabon. This area encompasses one of the last remaining undisturbed moist tropical forest ecosystems in the Congo Basin. Having received verbal support from the political leaders of the subregion for its agenda, the critical task facing environmental interest groups is to now turn words to action.

What is also evident in the Declaration is the intent of the Central African states to pursue the development of the sector. The support for increased processing within the sector, mentioned above as a key concern of the Cameroon government, was one of the points that the signatories as a group agreed to pursue. It is interesting to note that the term *conservation* is always accompanied by terms such as *sustainable management*, *utilization*, and *development* in the Declaration, and the term *protection* is used only once. When asked about his expectations for the summit, the Secretary General of the African Timber Organization, Paul Ngatse-Obala, stated his desire to see an end to the idea that conservation and forest exploitation were incongruent objectives.

Il est bien possible qu'on puisse exploiter durablement les forêts et les conserver durablement pour le bien-être des générations futures. [It is entirely possible to both exploit forests and sustain the resource for future generations] (**Cameroun Tribune**, Wednesday, March 17, 1999).

E. IITA Forest Margins Ecoregional Approach

Developing improved methods for managing natural resources and cropping systems that can conserve or improve the resource base has proven time and again to be a difficult task (Anderson and Byerlee 1995). The ecoregional approach was developed to overcome some of the inherent difficulties of effective natural resource management research and commodity development in heterogeneous environments (see box E.1).

In contrast to varietal innovations such as the high-yielding dwarf wheat and rice varieties of the Green Revolution, which were successful over a wide geographic area, the adoption of many NRM innovations is conditioned by the variability in Africa of biological, physical, and socioeconomic elements. Interactions across scales ranging from fields, landscapes, households, communities, watersheds, and regions further complicate NRM research. The spread of seed and planting material innovations also faces problems posed by a highly heterogeneous environment and the wide diversity of staple food crops across the region. The efficient supply of both NRM and seed innovations must find methodologies for addressing this widespread heterogeneity.

IITA and IRAD deliberately chose the benchmark to encompass a broad set of biophysical and socioeconomic factors (critical to defining research and development domains) in order to ensure the extrapolation of resource management results across the ecoregion. The forest margins benchmark covers a contiguous gradient of population and market access (figure E.1).

The benchmark was also chosen to encompass other active research institutes and the infrastructure conducive to bringing to bear the multidisciplinary scientific critical mass needed for the development of rapid innovation. Included among the research and development organizations active in the benchmark are the first two agricultural experiment stations established in Cameroon (at Nkoemvone in 1949 and Nkolbisson in 1954), the University of Yaoundé, ICRAF, CIFOR, ORSTOM, CIRAD, FAO, WWF, WCS, the World Bank, the national ministries including agriculture and scientific research, and various NGOs and farmer federations. The proximity of research and development institutes in and around Yaoundé provides rich opportunities for collaboration and exchange on a broad spectrum of issues. Yaoundé also provides the opportunity to study the impact of urban market access on traditional slash-and-burn systems.[7] The Humid Forest Experiment Station of IITA is centrally located in the Mbalmayo block. The station design uniquely supports the on-farm research in the benchmark villages.[8] Managed by researchers to simulate conditions found in farmers'

Box E.1. An Ecoregional Research Program for the Humid Forest

In the late 1980s, concerns were increasingly raised about the sustainability of agricultural production and its impacts on the environment. The challenge posed was to increase food output without degrading the natural resource base on which sustained production depends, while minimizing negative effects on environmental quality. In other terms, to develop a doubly green revolution. In pursuit of this goal, the Technical Advisory Committee of the Consultative Group on International Agricultural Research (CGIAR) proposed an ecoregional approach to focus and coordinate the research skills of the CGIAR (TAC 1993). The warm humid and sub-humid tropics in Sub-Saharan Africa were among the priority ecoregions. In April of 1996, the Ecoregional Programme for the Humid and Sub-humid Tropics of Sub-Saharan Africa (EPHTA) was launched at IITA, Ibadan, to pursue the challenge. The EPHTA program consists of three agro-ecosystem research consortia—the Inland Valley Consortium, the Moist Savanna Consortium, and the Humid Forest Consortium, comprised of advanced research institutes, international agricultural research centers, national agricultural research and extension services, nongovernmental organizations, and farmer organizations representing seven member countries—Cameroon, Côte d'Ivoire, the Democratic Republic of Congo, Gabon, Ghana, Guinea, and Nigeria. Following several planning meetings with Humid Forest Consortium member countries, research and development priorities were elaborated for the purpose of achieving improved livelihoods and competitiveness of forest zone producers and processors while protecting the forest ecology. Research within the consortium is currently concentrated in three benchmark areas—the Forest Pockets Benchmark located in southwest Ghana, the Degraded Forest Benchmark of southeast Nigeria, and the Forest Margins Benchmark of southern Cameroon.

Figure E1. The Forest Margins Benchmark and the Population Gradient, 1977

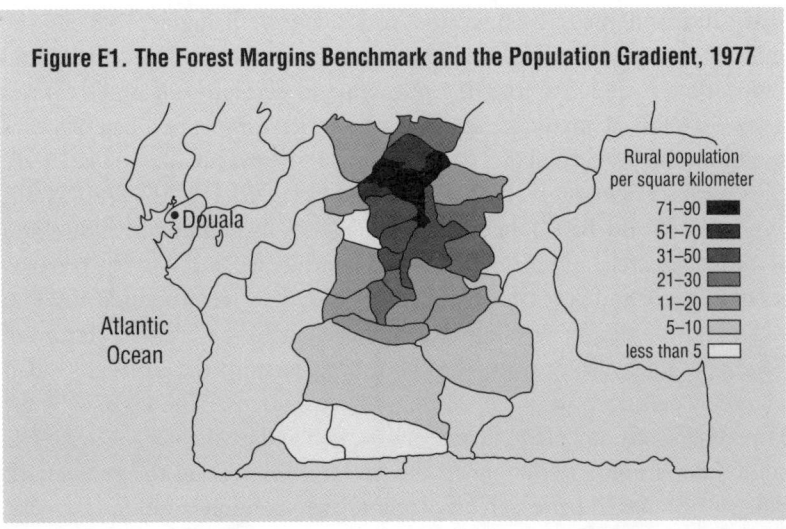

Rural population
per square kilometer

71–90
51–70
31–50
21–30
11–20
5–10
less than 5

Douala

Atlantic
Ocean

fields, the station focuses on understanding the fundamental processes associated with slash-and-burn agriculture and develops prototype models for on-farm adaptation.

Spanning the benchmark across a gradient of biophysical and socio-economic conditions allows the investigation of a continuum of research and development problems over a range of resource use intensities. Once an innovation is developed, the location-specific nature of demand requires a methodology for scaling-up for broader impact. Thus, characterizing and understanding the underlying patterns in heterogeneous environments is crucial both for the benchmark and the larger ecoregion. Natural resource management research must develop and target resource-conserving interventions to regions where the private benefits are most likely to be of sufficient magnitude to induce behavioral change. Identifying the underlying causal conditions under which this is likely to be the case and the spatial extent of those domains can greatly increase the environmental and economic impact of such research.

Once domains are identified, resource degradation and management problems are then diagnosed and prioritized. Characterization across the ecoregion is particularly important for setting broad research priorities. Once ecoregional priorities are identified, further in-depth diagnosis is conducted on the corresponding domain within the benchmark. Criteria for prioritization should include the importance of the problem, the spatial extent of the domain, and the likelihood of resolving the problem. After problem diagnosis and prioritization, the development of appropriate interventions (primarily in an on-farm, participatory, multidisciplinary approach) is undertaken within a similar domain in the benchmark.

Several approaches to characterization and domain definition are being developed for the benchmark. Macro-characterization of both the moist savanna and humid forest ecoregions has been conducted using secondary data sources and key informant interviews (Manyong et al. 1996a, 1996b). Another approach currently being pursued at ICRAF's Decision Systems Support unit is the development of multivariate benchmark similarity indices using a GIS of digital elevation, road and population densities, access to towns (markets), climatological data, and a normalized difference vegetation index (NDVI). Yet another approach is that of Walsh et al. (forthcoming), who have used a probabilistic Bayesian approach to model the risk of deforestation across the Congo Basin (including the benchmark). The model predicts where risk of deforestation due to slash-and-burn practices is highest, integrating remotely sensed fire data with spatial information on population densities and market access (as measured by distance to the nearest major town). Approaches such as these should be useful for targeting natural resource management interventions to appropriate domains for further adaptive testing and development.

Another characterization approach currently being implemented by EPHTA is a resource management survey administered on a uniform grid sampling basis. The goal of the survey is to define rural development patterns spatially. Typically, a higher resolution of characterization is necessary in the benchmark to better target interventions. For the EPHTA resource management survey, a sampling grid of 10x10 minutes is used in the benchmark, while a grid of 30x30 minutes is proposed for the broader ecoregion.

The linkages and complementarities among various characterization approaches should be exploited. The greater wealth of information available in a resource management survey for precisely defining development patterns (cropping systems, field management practices, purchased input use, revenue sources, etc.) must be weighed against its high cost of implementation. Correlations between pixels of a particular RMS development pattern with the data layers available in regional GISs could be exploited to determine extrapolation domains across a wider area at a much lower cost. Further development of GIS modeling capacity in the NARSs of the Congo Basin should be a priority for future training activities.

Within the benchmark, six villages have been chosen as representative of the range of resource use intensities likely to be found in the Congo Basin ecoregion. In these villages technological and management interventions are being developed and tested in both researcher- and farmer-managed trials. The six villages are located across the gradient of resource use intensities in the benchmark. The northernmost pilot village is Nkometou II in the Yaoundé block, which lies on the national road to the western provinces and has excellent access to the Yaoundé market. Nkometou falls under the administration of the Obala subdivision, which has the highest rural population density (72 persons/km^2) among the subdivisions encompassing the six villages. Nkolfoulou is even closer to Yaoundé, but lies within a subdivision (Soa) with lower population density (39 persons/km^2). This village is close enough to Yaoundé (12 km) that many of its population are employed in Yaoundé. Consequently, labor is a constraint because of its high opportunity cost. Both of these villages lie within the *Cultivation and Forest Regrowth Mosaic* land use of Millington and Pye (1994). In the Mbalmayo block, the village of Awae, south of Mfou, is characterized by relatively poor market access and moderate population pressures (37 persons/km²). Mvoutessi, lying on the paved road between Mbalmayo and Sangemelima, has good market access and abundant land resources (17 persons/km²). Both Awae and Mvoutessi fall within the Mesophilus Humid Tropical Forest land use category. In the southern Ebolowa block, Akok has abundant land resources (7 persons/km²), but is relatively inaccessible, with low vehicular traffic. In contrast, Mengomo, lying on the main Gabon-Equatorial Guinea trade route, has good access to the frontier market at Abang Minko'o, high vehicular traffic, and low population pressure (4 persons/km²). These latter two research villages fall into the Ombrophilous Humid Tropical Forest land category of Millington and Pye (1994).

F. ASB Evaluation of Land Use Systems in the Humid Forest Margins Benchmark of Southern Cameroon

The Alternatives to Slash and Burn (ASB) program of the CGIAR system has as its goal the development of technology interventions and policy incentives to support rural livelihoods in moist tropical forests. The program has conducted extensive diagnosis and characterization of land use systems in benchmarks in Indonesia, Brazil, and Cameroon, examining parameters associated with carbon sequestration, biodiversity, and socioeconomic performance. The following summarizes some of the chief findings from Cameroon to date.

The measure of carbon stocks developed by ASB for a given land use system is the average annual carbon stock measured in tons/ha over one cycle of the system. Mathematically, it is the time integral of the equation describing carbon sequestration divided by the number of years in the cycle (see box F.1). Above- and below-ground carbon was measured in six land use types: annual crop fields, short fallow fields of 1 to 4 years dominated by the woody herbaceous species *Chromolaena odorata*, medium fallow fields of 7 to 9 years, older fallow fields of 13 to 18 years, forest plots 50 to 100 years old; and mature cocoa perennial crop systems 25 years old. The Ruthenberg ratio, $r = t_{crop}/(T_f + t_{crop})$, where t_{crop} is the period of cropping and T_f is the fallow period, is a useful descriptive index for cropping intensity in fallow rotational systems (Ruthenberg). As r approaches 1, the intensification of cropping over time increases until continuous cultivation when $r = 1$.

Total time-averaged carbon stocks for the four principal administrative divisions of the benchmark and for the various landuse systems are reported in table F.1.[9] In terms of agricultural intensification, these divisions can be ranked as follows: Lékié > Mefou > Nyong et So'o > Ntem. Overall, we estimate that the effects of agriculture have reduced the total carbon pool in the benchmark to 80 percent of its estimated primordial level (table F.1). Carbon stocks decline with agricultural intensification, declining from 89 percent of the primordial level in the Ntem division to 36 percent in the intensely cultivated Lékié division. The share of total carbon in perennial crop systems increases with overall intensification. In the Lékié division, it is estimated that there are 2.5 million tons more carbon in cocoa agroforests than in the remaining forested land. This is an extremely important finding, for it shows that the shaded cocoa and coffee agroforests practiced by smallholders of southern Cameroon offer an intermediate step between the low carbon, short fallow systems and primary forest.

Box F.1. Time-Averaged Carbon Stocks

The data for carbon pools in various natural fallow cover types as reported in Kotto-Same et al. were used to estimate the empirical relationship between time and total system carbon (equations 1 and 2). The functional relationship between fallow length and carbon was combined with that for the cropping phase to determine the overall carbon-time relation, $f(x_{cf})$ during the crop-fallow rotation:

$$f(x_{cf}) = 79.8 \qquad \text{for } -t_{crop} \leq x_{cf} < 0 \qquad (1)$$
$$= 79.8 + 4.90\ x_{cf} + 1.25\ x_{cf}^2 - 0.065\ x_{cf}^3 \text{ for } 0 \leq x_{cf} \leq T$$

where x_{cf} are years in the crop-fallow cycle, t_{crop} are years cropped, year 0 is the start of the fallow period, and T are the years of fallow.

For perennial crop systems common to southern Cameroon (shaded cocoa, shaded coffee, and fruit-tree-based home gardens), the estimated functional form $f(x_p)$ was linear:

$$f(x_p) = 79.8 + 4.216\ x_p \qquad \text{for } 0 \leq x_p \leq 25 \qquad (2)$$
$$= 185 \qquad\qquad \text{for } 25 < x_p < T$$

where x_p are years in the perennial crop system, year 0 is the start of the establishment period, and T is the age of the plantation. As all perennial crop measurements were taken in 25-year-old plantations, there are no regression properties associated with this function, which is simply the line fit between mean carbon estimates for $T = 0$ and $T = 25$. Plantations older than 25 years were assumed to be in a steady state at 185 tons ha[-1]. Forested land is also assumed to be in a carbon steady state at 307 tons/ha.

The time-averaged carbon stock (LUS_C) for annual and biennal crop-fallow rotational systems is the mean value of the integral of $f(x_{cf})$ over the interval $[-t_{crop}, T]$:

$$LUS_C = (\int f(x)\ d_x)/(T + t_{crop}) \qquad (3)$$

Table F1. Carbon Stocks Across Land Use Categories and Administrative Divisions of the Forest Margins Benchmark Area of Southern Cameroon (millions of tons, Mt)

Land use system	Lékié Mt	Lékié % of total	Mefou Mt	Mefou % of total	Nyong et So'o Mt	Nyong et So'o % of total	Ntem Mt	Ntem % of total	Subtotal Mt	Subtotal % of total
Current levels	33.2	100	107.7	100	94.5	100	435.0	100	667.7	100
Annual food	16.5	50	10.0	9	4.3	5	12.9	3	42.8	6
Biennal food	6.2	19	8.3	8	4.4	5	25.4	6	42.3	6
Cocoa	6.8	21	4.1	4	1.9	2	6.0	1	19.0	3
Forested land	3.7	11	85.3	79	83.9	89	390.7	90	563.6	84
Primordial levels	91.8		144.7		109.9		491.2		837.4	
Current levels as % of primordial	36%		74%		86%		89%		80%	
Average C/ha (t)	111 t		229 t		264 t		272 t		245 t	

Source: Gockowski et al. 1998.

A 1994 ASB survey conducted in the four administrative divisions of the Forest Margins Benchmark estimated that 86 percent of crop lands are made up of variants of three cropping systems: cocoa, groundnut-cassava based intercrops, and plantain-cocoyam-*Cucumeropsis manni* based intercrops (Gockowski et al. 1998). In all, 24.8 percent of the total land area in these divisions was in agricultural use. The most widespread system was the cocoa agroforest, accounting for 115,000 ha or 48 percent of all cropland followed by the plantain-cocoyam-*Cucumeropsis manni*-based system at 21 percent, and the mixed groundnut field at 17 percent. The latter two are annual/biennal cropping systems that convert fallow to crop land via slash-and-burn techniques.

The photochemical processes of cocoa agroforests capture solar radiation more efficiently than annual food crop systems, which through human manipulation is turned into useful outputs. The mid- and upper strata provide shade; recycle nutrients; maintain plant, insect, and microbial diversity; and yield many useful products. Indigenous fruit trees such as the African plum (*Dacryodes edulis*), bush mango (*Irvingia gabonensis*), and njansang (*Ricinodendron heudelotii*), along with exotics such as citrus, mango, guava, and avocado make an important contribution to local diets and provide additional revenue sources for many households.

Commercial timber species provide lumber used in local rural construction, while oil palm (*Elaeis guineensis*) provides cooking oil, wine, and distilled spirits. Lower strata include useful vines such as *Gnetum* spp. whose leaves are highly prized as a leafy vegetable dish in many parts of Cameroon, Nigeria, Gabon, and the Congos. Many medicinal plants are also maintained, including the malaria-suppressing *Alstonia boonei*. In areas of high population pressure such as the Lékié division just to the north of Yaoundé, African gray parrots and hornbills still thrive, thanks to the diverse habitat and abundant fruit supply provided by cocoa agroforests.

Cocoa agroforests generate significant economies, external to the household, that benefit society at large. These include the maintenance of the local hydrological cycle and local precipitation patterns, control of soil erosion, and the conservation of important genetic resources, particularly for indigenous fruit trees. They also store more carbon than annual crop systems, attaining up to 60 percent of the carbon stock of the original forest (ASB 2000). In sum, the cocoa-fruit agroforests of southern Cameroon provide a solid foundation for the maintenance of ecosystem stability and should be the focus of directed international, national, provincial, and local policies to maintain these economically and environmentally important systems on the landscape.

In addition to its environmental advantages, the cocoa agroforest offers a broad-based and equitable rural development pathway. In the Center and South Provinces of southern Cameroon, approximately 75 percent of rural households produce cocoa and the median agroforest is 0.87 ha (MINAGRI 1988). These land use systems, unlike timber exploitation or large-scale rubber and oil palm plantations, are usually accessible to even resource-constrained rural households. Unfortunately, revenues in the last 20 years from these systems have steadily eroded because of weakness in world markets as well as policy and institutional factors, with negative consequences for the welfare of cocoa producing households and the environment.

The carbon stock in crop-fallow rotational systems is an indicator of intensification and system sustainability. In moist tropical forests the ratio of nutrients in biomass to that in the soil is far higher that that in temperate forests.[10] Total carbon is proportional to total biomass, while total biomass is positively correlated with total nutrient stocks.[11] The carbon stock is thus an approximate indicator of the total nutrient stock of a land-use system. If carbon levels are high, the elevated quantity of nutrient-rich ash following the burn will produce sufficient returns to labor to maintain rural livelihoods and reproduce the system.

However, once the system carbon is run down and fertility reduced, intensification in the form of new management techniques, improved planted fallow, new crops, purchased inputs, and new cropping sequences and rotations are required in order to sustain rural livelihoods. The average carbon stock in a crop-fallow rotation system with a Ruthenberg ratio of 0.5 is estimated to be 45 percent lower than an extensive system with a value of 0.11 (table F.2). If the nutrient composition were the same for the two different fallow periods, total nutrients in the intensive system at the time of slashing, T, would be 58 percent lower than in the extensive system. Rural households in the Lékié division of the Yaoundé block are most likely to have Ruthenberg ratios approaching or exceeding 0.5. These households have intensified the field management of their food crop systems to a much greater degree than other areas of the benchmark.

The carbon stocks sequestered in cocoa agroforests may help to reduce global warming and should be considered in discussions of carbon sinks and emissions trading (Newmark). The conversion of one hectare of a short fallow-crop rotational land use to a cocoa agroforest could sequester up to 95 tons of carbon, depending on the fallow length and the cocoa production cycle. Cocoa agroforests also can play an important role in maintaining hydrological functions. Deforestation can dis-

Table F.2. Dynamics of Carbon Stocks for Various Land Use Systems in Southern Cameroon

	Crop period /age of tree stock (years)	Fallow period (years)	Total carbon at slashing (t C/ha)	Time-average carbon (t C/ha)	Ruthen-berg ratio
Short fallow intercrop	1.5	1.5	90	82	0.50
Short fallow intercrop	1.5	3.0	104	87	0.33
Short fallow intercrop	1.5	4.0	115	91	0.27
Medium fallow intercrop	1.5	5.0	127	96	0.23
Medium fallow intercrop	1.5	7.0	153	106	0.18
Medium fallow intercrop	1.5	9.0	178	117	0.14
Long fallow intercrop	2.0	12.0	206	132	0.14
Long fallow intercrop	2.0	14.0	215	142	0.13
Long fallow intercrop	2.0	16.0	212	150	0.11
Cocoa plantation	15.0	—	143	111	—
Cocoa plantation	25.0	—	190	132	—
Cocoa plantation	40.0	—	190	154	—
Primary forest	—	—	307	307	—

Source: Gockowski et al. 1998

rupt the energy/water balance, local precipitation patterns, drainage, runoff, and water yield. The scale at which these effects occur is not easily discerned because of the difficulty in getting before and after data for large areas. The relatively large area still in perennial crop systems in the Lékié division (an estimated 13 percent of total area) might help to maintain the hydrological functioning of the environment and precipitation patterns.

The environmental importance of maintaining cocoa agroforests on the landscape has perhaps not been fully recognized by national and international decisionmakers. Evidence for this opinion is found in policies undertaken in the name of structural adjustment, which severely undermined cocoa producer incentives in the late 1980s and early 1990s.[12]

The Effects of Intensification on Plant Biodiversity

ASB only measured plant diversity at the plot level. The biodiversity measure used by ASB incorporates the notion of plant functional attributes developed by Andy Gillason of CIFOR with standard measures of species richness. Six plant functional attribute classes are used—leaf size, leaf inclination, chlorotype, leaf type, life form, and root type. Each attribute class is composed of a subset of elements. For instance, the three possible elements for the leaf inclination class are vertical, lateral, pendulous, and composite. The combination of all classes and their descriptive elements is referred to as the plant modus. Each species will be associated with one modus. The ratio of species to plant modi measured in a 40 x 5 m transect is used as an indicator of plant diversity. This ratio, by combining the more commonly used measure of species richness with plant functions, provides a measure of the robustness of the ecological functions of the system. Land uses with a higher ratio (i.e., more species for a given number of modi) would be expected to have greater ecological resiliency in the face of change. Biodiversity in crop-fallow rotational systems was evaluated by comparing points in the time sequence having the highest species:modi ratio. The initial phase of ASB did not study the relationships between these measures of plant biodiversity and the biophysical functioning of agricultural land use systems.

Not surprisingly, biodiversity declines as the fallow-period shortens (table F.3). Among agricultural systems, the species:modi ratio and species richness were highest for the cocoa agroforest, representing 80 percent of the forest measures. The importance of the biodiversity in this land use system grows as population pressures build. In the intensely cropped Lékié division, only 20 percent of the landscape still meets or exceeds these levels. However, of that 20 percent, two-thirds is accounted for by cocoa agroforests, with the remaining one-third found in the few forest remnants still existing.

The cocoa agroforest is particularly effective in maintaining the biodiversity most valued by rural populations. These complex multistrata systems serve as biological reserves for many forest products used and traded by rural populations. Over

Table F.3. Measures of Plant Biodiversity by Land Cover Type in Southern Cameroon

Land cover	Modi	Species richness	Species: modi ratio
Raffia forest	29	57	1.97
Secondary forest	39	76	1.95
Cocoa plantation	38	63	1.66
Bush fallow (8–15 years)	35	53	1.51
Chromolaena fallow (2–4 years)	44	64	1.45
Annual food crop field	32	42	1.31
Forest food crop field	12	14	1.17

Source: ASB—Cameroon Phase II Final Report 1999.

the last 60 years, where land pressures have increased and the forest has largely disappeared, farmers have nurtured and transplanted wild seedlings of indigenous fruit and timber trees into their cocoa agroforests. In the subdivision of Sa'a, one of the most heavily populated areas in the benchmark, the production and marketing of the widely consumed oilseed *"ndjansan"* from the forest tree species *Ricinodendrom heudelotii*, is an important commercial activity. Most of this production originates from cocoa agroforests where this species also serves as shade for the cocoa understory.

While the international community is greatly concerned with the maintenance of the environmental services discussed above, poor farmers eking out a marginal existence can only be concerned with the capacity of their farming systems to maintain livelihoods. We now examine the economic performance of the major land-use systems in the benchmark and their contribution to household liquidity and sustenance.

Comparative Economic Analysis of the Various Land Use Systems

Policy analysis matrices (PAMs) were constructed to evaluate the economic returns and policy distortions affecting the major land use systems and hybrid oil palm production systems (Monke and Pearson). The net present value (NPV) of costs and returns over a 30-year period was calculated using a 10 percent discount rate and an opportunity cost of family labor of US$1.21 per person day. Six perennial crop systems were evaluated: (1) Intensive cocoa with mixed fruit tree shade canopy planted into short fallow; (2) Intensive cocoa planted into short fallow; (3) Extensive cocoa with mixed fruit tree shade canopy, planted into forestland/long fallow; (4) Extensive cocoa planted into forestland or long fallow; (5) Improved Tenera hybrid oil palm system planted into short fallow; and (6) Improved Tenera hybrid oil palm system planted into forested land or long fallow. These were compared to the two dominant slash-and-burn food-cropping systems in the benchmark: (1) Intercropped groundnut/cassava-based food field planted into a short fallow; and (2) Intercropped melon/plantain/cocoyam food field planted into a long fallow.

Together these last two cropping systems and the cocoa agroforest account for an estimated 87 percent of all cropland in the benchmark area (Gockowski et al.). The four variants of cocoa agroforest reflect differences in the intensification and commercialization of these systems across the benchmark. As mentioned above, cocoa agroforests are more intensively managed in the Yaoundé block than the extensive systems found elsewhere in the benchmark. A distinction is also drawn between agroforests with urban market access for secondary fruit tree production and those without a commercial fruit tree component.

The domestic resource cost ratio was less than one for all systems (table F.4).[13] The intensive cocoa with fruit and the hybrid oil palm system in forested land had the lowest domestic resource costs (DRCs) (0.49 and 0.48), while the extensive plantain-based system, extensive cocoa system without fruit, and the short fallow rotation had the highest. Effective protection coefficients (measuring the degree of taxation or subsidy) for the perennial crop systems were in the range of 0.89 to 0.93.[14] Taxation on cocoa (mainly consisting of import tariffs on pesticides and a 10 percent excise tax on production) is lower today compared to when the national marketing board was operating and setting producer prices by presidential decree. The liberalization of export markets in 1994, particularly now that world market prices have picked up, is providing greater incentive to cocoa and coffee producers. Internal food crop markets are not distorted as indicated by the effective protection coefficients (EPCs) of 1.0 for the rotational fallow systems.

The return to labor is a critical variable for determining the adoption potential of a given technology system, particularly in land-abundant, labor-scarce economies such as that of the Ebolowa block of the benchmark. A grid search was conducted to determine the wage rate at which the NPV was equal to zero. This was used as the measure of the returns to labor. The intensive fruit-cocoa system and the oil palm system planted

Table F.4. Selected PAM and Environmental Results for ASB Land Use Systems

	Domestic resource cost	Effective protection coefficient	Social profit-ability (US$/ha)	Private returns to labor (US$/day)	Time-averaged carbon stock (t C/ha)	Bio-diversity (species: modi ratio)
Int cocoa w/fruit	0.49	0.90	1,755	2.36	154	1.66
Int cocoa	0.58	0.89	1,236	1.95	154	1.66
Ext cocoa w/fruit	0.68	0.90	1,136	2.13	154	1.66
Ext cocoa	0.54	0.93	616	1.63	154	1.66
Sf oil palm	0.62	0.90	982	1.81	153	1.18
For oil palm	0.48	0.92	1,654	2.44	153	1.18
Sf intercrop	0.69	1.00	644	1.79	82	1.45
Lf intercrop	0.73	1.00	288	1.70	142	1.51

Source: Gockowski et al. 1998.

into forestland had the highest returns (US$2.36 and US$2.44 per person day respectively), and the extensive cocoa and long fallow plantain systems the lowest (table F.4).

Gockowski and Dury (1999) further investigated differentiation in cocoa agroforests based on ongoing IITA fieldwork in the Lékié division. The extent of current policy distortions for five typologies of cocoa agroforests are presented in table F.5 and show that distortions are higher, the less diversified the system (into fruit trees). The establishment of new cocoa-fruit agroforests should be encouraged. However, poverty can negatively impact the adoption rate of new technologies, particularly those with a substantial biological time lag before production. Poor farmers in the humid forest zone of Indonesia facing liquidity constraints and with limited resources were shown to have short planning horizons, to be more risk-averse, and with little option of waiting for long-term investments in tree stock to yield fruit (Dury 1997).

The issue in Cameroon was examined by varying the discount rates from 8 to 25 percent. At the lower discount rate, net present value was estimated to be US$3,100/ha and the perennial component of the agroforest accounted for 90 percent of total discounted gross revenue. At the highest discount rate (representing risk-averse, liquidity-constrained poor farmers), net present value was US$100/ha and annual crops grown during the establishment phase of the plantation contributed 40 percent of total gross revenues. This highlights the importance of addressing the land and labor productivity of annual crops during the establishment phase if poverty alleviation and the participation of the poor are among the targeted objectives.

Table F.5. Policy Distortions and Returns to Land and Labor of Cocoa-Fruit Agroforests in Southern Cameroon

	Returns to labor (US$/day)	Net present value (US$/ha)	Domestic resource cost ratio[a]	Subsidy ratio to producers	Effective protection coefficient
Medium intensity cocoa-fruit agroforest	$2.36	$1,755	0.49	-7.73%	0.90
High intensity cocoa-fruit agroforest	$3.60	$3,687	0.56	-6.49%	0.95
High diversity "Obala" agroforest	$3.27	$2,605	0.38	-4.22%	0.95
Extensive cocoa-fruit agroforest	$2.14	$943	0.54	-6.31%	0.93
Extensive cocoa	$1.63	$424	0.68	-7.67%	0.90

a. Domestic resource cost = cost of domestic factors of production divided by returns to domestic factors of production.
Source: Gockowski and Dury 1999.

G. The Political Economy of Agricultural Liberalization

Ntangsi (1987) classifies Cameroonian interest groups affecting agricultural policy into five categories that are useful for a discussion of the political-economic influences on price policy in southern Cameroon. These categories as they pertain to the coffee and cocoa sectors are: (1) ethnic-regional groupings (the Bamiliké and Tikar coffee producers of the western provinces, and the Beti-Boulou-Fang cocoa producers of southern Cameroon); (2) the peasantry (accounting for approximately 100 percent of cocoa and robusta coffee production); (3) institutional groups (the Presidency; political parties; the NPMB, MINGRI, MINDIC bureaucracies); (4) urban-based elites (including marketing and processing agents, exporters, along with bureaucratic elites); and (5) the external aid donors (the World Bank, IMF, *Caisse Centrale*, EDF, STABEX, and USAID).

Peasantry, Ethnicity, and Political Parties

From the late 1950s through the mid-1960s the *Union des Populations du Cameroun* (UPC) waged a guerilla war against the colonial and post-independence Ahidjo regime (backed by the ex-colonial power). The underground movement originated in the Nyong et Kelle region of the Center Province, led by Reuben Um Nyobé. After Nyobé was killed by French troops in 1959, the locus of the insurgency shifted to the western provinces. Adhijo finally squashed the movement with the capture of Earnest Ouandji in Kumba, Southwest Province, in 1970. As a carrot to the population of West province (which was a stronghold of the UPC), Ahidjo granted exclusive purchase and export selling rights to UCCAO, the arabica cooperative union. Allowing UCCAO to sell and retain the profits from arabica coffee was a major departure from the way export crops were marketed in all other regions of the country and allowed UCCAO and its cooperative members to prosper.

A change in the regional distribution of power occurred when President Ahidjo from the north was succeeded by Paul Biya of the Beti-Bulu Center-South in 1982. Biya maintained the one-party system of Ahidjo until November of 1990, when pressures from other political parties forced the change to a multiparty political system.[15] The major opposition to RDPC came from the populist Social Democratic Front (SDF) founded in Bamenda in 1990, with most of its power base in the coffee growing regions of the western provinces. In 1992, a very close presidential election result was adjudicated by the Supreme Court, in favor of the incumbent Paul Biya of the RDPC. Biya's victory was in large part due to the majority backing of the cocoa growing Center and South.

Biya rewarded his political supporters among Beti-Bulu cocoa growers by maintaining the official cocoa producer price at its 1991/92 level, despite continued downward pressure on world prices, whereas coffee prices were again lowered in late January of 1993.

This change in political institutions (from a single party to a multi-party system) seems to have invested more political power in the peasantry than was previously the case. Also contributing to an increasing political voice for farmers was a 1991 law encouraging the formation of farmer organizations as a means of counterbalancing the perceived threat to cocoa and coffee producers in a newly liberalized free market.

Prior to the move toward a multiparty system, Ntangsi wrote (p. 42):

The Cameroon peasantry are not directly represented in the political system, do not possess formal political power and therefore do not directly influence policy-making.

While their political influence is still slight vis-à-vis urban-based groups, continued movement toward a multiparty system and the increasing role played by farmer organizations will continue to invest the peasantry with political power.

Urban Based Elites: Marketing Middlemen and Exporters

The economic fortunes of urban-based marketing middlemen and exporters were determined from independence until liberalization by government regulation. Approximately 150 licensed buying agents operated in the cocoa and coffee sectors prior to liberalization in 1991. These agents collected and treated the producer's cocoa and coffee. In addition to these middlemen, there were approximately 60 licensed exporters of robusta coffee and cocoa. Until the recent reforms (see below), these exporters historically paid a fixed price to the licensed buying agents for the producer's hulled green coffee and dried and fermented cocoa. They in turn graded, sorted, and contracted sales in the world market, all at prices fixed by the national marketing board (ONCPB). Before it was liquidated in 1991, the ONCPB also acted as an exporter of coffee produced in the anglophone areas of Cameroon (i.e., the Northwest and Southwest Provinces). On an individual basis, the exporters have unequivocally the largest economic stake of any of the actors in the coffee and cocoa sectors.

Bureaucratic Interest Groups and the Coffee and Cocoa Sector

Bureaucrats affect economic policies and relative prices in the cocoa and coffee sectors in several ways: (1) indirectly, the significant wage bill of the civil service affects agricultural terms of trade; (2) via their direct influence on policy formulation through policy analyses; and (3) through their implementation of government policies.

As with any bureaucratic structure, opportunities for rent-seeking behavior arise and can shape the behavior of the bureaucratic agent. Bureaucratic rent-seeking and government acting as Leviathan, seeking to maximize its legitimacy by enlarging the public sector, has been developed by Findlay and Wellisz (1983). Tariffs are set to maximize revenues, which are then used to enlarge the state sector. This idea of the autonomous state has been extended to explain the logic of marketing boards as a means of expanding state bureaucracy (Findlay 1991). Their theoretical modeling seems to find some empirical content in the care of the Cameroon marketing board.

At independence, in both West Cameroon and East Cameroon, the new federal government inherited British and French state marketing organizations of differing natures. In the former British protectorate of West Cameroon, the National Marketing Board (NMB) undertook physical market functions including collection, processing, conditioning, and export of robusta and arabica coffee and cocoa. In the former French protectorate, the *Caisse de Stabilisation* oversaw the commercialization of the smallholder robusta coffee and cocoa, but in contrast to the NMB, never undertook any of the physical exchange functions. Instead it used the above-mentioned licensed marketing agents to collect, process, and export the beans in exchange for a fixed commission determined annually by the *bareme* (a set of administered official prices for the coffee and cocoa sectors). In essence, the *bareme* established the price for every marketing function. The *bareme* was annually issued by presidential decree after the recommendations to the presidency by the marketing board and the Ministry of Industry and Commerce.[16] In 1975, these two institutions were united into a single office, the *Office National de Commercialization des Produits de Base* (ONCPB), although functionally nothing changed in the way coffee and cocoa were exported from the two regions.

Over the two identifiable booms in the coffee and cocoa markets since 1975, the ONCPB generated positive surpluses. In principal, these surpluses were to be retained for years when world prices were low. However after the collapse of oil prices in early 1986 and given the growth in the public sector and budget discussed above, the government rapidly turned to ONCPB surpluses that had been generated by the high world prices for coffee and cocoa in the mid 1980s. In 1986–87, the financial woes of the

ONCPB grew rapidly as cocoa prices fell, followed in 1987–88 by coffee prices. The continued slide in commodity prices and the exhaustion of the ONCPB stabilization funds resulted in widespread payment arrears in 1988–89 to both marketing agents and producers in the cocoa and coffee sectors (over US$300 million).

With the government budget in serious deficit and the ONCPB on the verge of bankruptcy, the last of Ntangsi's stakeholders—the external donors—began to play a significant role in the economic regulation of the coffee and coca sectors.

External Donors, the Bureaucracy, and Liberalization

In 1988, Cameroon began negotiations for the first time in its young history with the World Bank and IMF on a Standby Arrangement, a structural adjustment loan (SAL), and a policy reform program. In 1989, a SAL for US$150 million and a reform program were agreed upon, with the SAL to be disbursed in three tranches. As part of the program, it was agreed that a financial audit of the ONCPB was required to guide its restructuring. Based on the results of the audit, the World Bank recommended the dissolution and liquidation of the ONCPB, whose work force had grown to over 2,800 employees. The Presidency of the Republic was opposed, and in September of 1990 the World Bank froze the second tranche of the SAL, while the IMF suspended Cameroon's Standby Arrangement, citing the slow implementation of the agreed upon policy reform package.

With approximately 100 billion CFA francs required to settle all its outstanding debts to producers and market agents from the 1988–89 campaign, the government of Cameroon turned to a coalition of European donors: the French *Caisse Centrale de Coopération Economique* and the European Economic Community. Funding was provided in 1990 by the Caisse Centrale and STABEX funds from the European Development Fund in the amount of 52.2 billion CFA francs, which was used to settle a large part of the arrears from the 1988–89 season. This injection of liquidity allowed the 1989–90 campaign to be carried out following a major restructuring of the ONCPB. Its work force was reduced from over 2,800 to 1,100 and its annual operating budget cut from 12.5 billion to 5 billion CFA francs. The *bareme*, commonly referred to as "*les charges incompressibles*" (irreducible costs), were significantly reduced for 1989–90. In 1990–91, with continued falls in coffee and cocoa markets and the external donors creating pressure for further reforms, the ONCPB was dissolved and replaced by the *Office National du Café et*

Cacao (ONCC) with a staff of 157. The exporters organized themselves into the *Groupement des Exportateurs* (GEX), in a lobbying response to the continuing compression of the bareme.

The newly created ONCC, no longer involved in physical market activities, was only charged with regulating a reduced set of fixed prices and quality control. The government of Cameroon still officially set the producer price of robusta and the c.i.f. price of the exporter (in negotiation with GEX), but all other prices were allowed to adjust freely according to internal supply and demand for the marketing services of the licensed buying agents. If the world price was higher than the c.i.f price set by government, these revenues were to be paid into a stabilization fund at the *Banque des Etats d'Afrique Centrale* rather than the national treasury, as was previously the case. Finally, in 1995, the state completely withdrew from price fixing in the cocoa and coffee sectors and prices were freely determined.

H. Shift in World Bank Strategy in Cameroon

Sectoral Focus

The World Bank has been engaged in Cameroon since 1967. Prior to the 1994 devaluation, there was a noticeable drop in lending due largely to poor country performance. From the 1984–91 period to the 1992–99 period total commitment to Cameroon increased by about 11 percent, from US$733.3 million to US$817 million. The total number of projects also increased from 14 to 18. However, average commitment per project decreased from US$52.38 million to about US$45.39 million. In addition, the structural composition of the portfolio changed dramatically.

Tables H.1 and H.2 clearly show a shift in sectoral focus from *agriculture*, with 50 percent of the projects and 31.6 percent of commitments in the first period, to *multisector*[17] activities and public sector management in the second. These activities accounted for more that 61 percent of the projects and more than 78 percent of the commitments in 1992–99. *Given poor country performance, it probably made sense to shift emphasis to activities that would build institutions and improve economic management.*

Instrumental Shift

Tables H.3 and H.4 also reveal a shift in the instruments used to implement the modification in the strategy. In the 1984–91 period, resources were channeled overwhelmingly through specific investment loans. These instruments (specific investment loans and specific investment and maintenance) accounted for 78.58 percent of the projects and about 75 percent of the commitments in 1984–91 versus 28 percent and 20 percent respectively in 1992–99. At the same time, the weight of structural adjustment operations increased from 7.14 percent of projects and 20.46 percent of commitments to 55.56 percent and 76.56 percent respectively. This is the instrument that the Bank decided to use to deal with forest sector issues in Cameroon starting in 1989. That is still the strategy. Indeed, the 1996 CAS explicitly stated: "to help the government consolidate the benefits of devaluation, and in close collaboration with the IMF, IDA would provide quick-disbursing lending through two adjustment credits in FY96 and FY97."

Deteriorating Country Performance

Table H.5 presents an overall assessment of the performance of the lending portfolio in Cameroon over the past two decades. From the 1980–91 period to the 1992–98 period, the proportion of projects with a satis-

Table H.1. Cameroon 1984–91: Sectoral Distribution of Projects

Sector	Number of projects	Percentage of projects	Commitment (US$M)	% of total commitment
Agriculture	7	50.00	231.7	31.60
Education	1	7.14	30.1	4.10
Multisector	1	7.14	150.0	20.46
Pop., Health, and Nutrition				
Public Sector MGT	1	7.14	9.0	1.23
Social	1	7.14	21.5	2.93
Transportation	2	14.29	145.0	19.77
Urban Development	1	7.14	146.0	19.91
Total	14	100.00	733.3	100.00

Source: MIS.

Table H.2. Cameroon 1992–99: Sectoral Distribution of Projects

Sector	Number of projects	Percentage of projects	Commitment (US$M)	% of total commitment
Agriculture	2	11.11	38.1	4.66
Education	1	5.56	4.9	0.60
Multisector	6	33.33	401.9	49.19
Pop. Health and Nutrition	1	5.56	43.0	5.26
Public Sector MGT	5	27.78	236.2	28.91
Social				
Transportation	3	16.67	92.9	11.37
Urban Development				
Total	18	100.00	817.0	100.00

Source: MIS.

factory outcome dropped significantly, from 61 percent to about 15 percent. Satisfactory projects accounted for about 74 percent of net commitments in the first period and only 17 percent in the second period. With respect to sustainability, table H.6 indicates that the percentage of projects whose outcome was rated likely to be sustainable dropped from 13 percent to zero. In the 1980–91, the projects whose outcomes was rated likely sustainable accounted for about 18 percent of net commitments.

Table H.3. Cameroon 1984–91: Distribution of Projects by Instruments

Instrument	Number of projects	Percentage of projects	Commitment (US$M)	% of total commitment
Financial Intermediary Loan	1	7.14	25.50	3.48
Learning and Innovation Loan	0	0.00	0.00	0.00
Sectoral Adjustment Loan	0	0.00	0.00	0.00
Structural Adjustment Loan	1	7.14	150.00	20.46
Specific Investment Loan	9	64.29	320.80	43.75
Specific Investment & Maintenance	2	14.29	228.00	31.09
Technical Assistance Loan	1	7.14	9.00	1.23
Total	14	100.00	1,052.10	100.00

Source: MIS.

Table H.4. Cameroon 1992–99: Distribution of Projects by Instruments

Instrument	Number of projects	Percentage of projects	Commitment (US$M)	% of total commitment
Financial Intermediary Loan	0	0.00	0.00	0.00
Learning and Innovation Loan	1	5.56	4.9	0.60
Sectoral Adjustment Loan	3	16.67	193.4	23.67
Structural Adjustment Loan	7	38.89	432.10	52.89
Specific Investment Loan	5	27.78	163.80	20.05
Specific Investment & Maintenance	0	0.00	0.00	0.00
Technical Assistance Loan	2	11.11	22.80	2.79
Total	18	100.00	817.00	100.00

Source: MIS.

To begin to understand the reasons behind such a poor performance, which defines the context in which any intervention for the forest sector would take place, we turn first to Country Portfolio Performance Reviews (CPPRs), conducted jointly by the government of Cameroon and the World Bank. Three such reviews were done between 1993 and 1998 (the second one took place from January 20 to 24, 1997). These reviews acknowledge the deterioration of the World Bank lending program in Cameroon and they point to the following key factors: (1) *poor*

Table H.5. Cameroon 1980–98 Aggregate Lending Outcome

	1980–91	1992–98
Satisfactory outcomes (SO)	19	2
Satisfactory outcomes (%)	61.29	15.38
Net commitments with SO	747.25	139.75
Net commitments with SO(%)	74.21	17.45
Unsatisfactory outcomes (UO)	12	11
Unsatisfactory outcomes (%)	38.71	84.62
Net commitments with UO	259.64	661.27
Net commitments with UO(%)	25.79	82.55
Total number of projects	31	13
Total net commitments ($m)	1,006.89	801.02

Source: OED evaluations.

Table H.6. Cameroon 1980–98 Outcome Sustainability Assessment

	1980–91	1992–98
Not-rated projects	15	0
Not-rated projects (%)	48.39	0.00
Not-rated net commitments ($m)	411.08	0.00
Not-rated net commitments (%)	40.83	0.00
Likely (projects)	4	0
Likely (% projects)	12.90	0.00
Likely (net commitments)	178.9	0.00
Likely (% net commitments)	17.77	0.00
Uncertain (projects)	7	4
Uncertain (% projects)	22.58	30.77
Uncertain (net commitments)	303.59	177.33
Uncertain (% net commitments)	30.15	22.14
Unlikely (projects)	5	9
Unlikely (% projects)	16.13	69.23
Unlikely (net commitments)	113.32	623.69
Unlikely (% net commitments)	11.25	77.86
Total number of projects	31	13
Total net commitments	1,006.89	801.02

Source: OED evaluations.

quality of the dialogue; (2) *poor management of both the economy and projects*; and (3) *weak institutions*.

The overall theme of the 1997 CPPR related to the need to improve *mutual trust and respect* between Cameroon and the World Bank. This seems to indicate a breakdown in partnership between the country and the Bank. The complaints from the Cameroon side relate to the following: (1) *the Bank tends to micromanage its relationship with Cameroon*; (2) *the conditionalities imposed are rather stringent, shifting and not adapted to country conditions*; (3) *intransigence in negotiations and project supervision*; and (4) *lack of mastery of local institutions*.

At that same occasion, the Bank pointed out that the government of Cameroon needed to improve its record on economic management and project implementation. Two key factors undermining project implementation in Cameroon are the inadequacy of counterpart funds and financial mismanagement at the project level. Counterpart funds are not enough and are disbursed late. For FY97, the necessary level of counterpart funding was estimated at about 25 billion CFA francs. Less than 2 billion was disbursed. As far as mismanagement is concerned, there have been cases where the Bank had to stop disbursement and ask the government to reimburse unjustified expenditures.

Endnotes

Summary

1. Projects classified as multisector relate to economic management, while public sector management involves institutional development and public enterprise reform.

Chapter 1

2. Appropriateness is defined here in terms of social costs and benefits using broad socioeconomic criteria. Also, it is useful to keep in mind that there may be a continuum of land use choices for any given area of forest, ranging from the one that would leave it undisturbed to those that imply clearing the forest and putting the land to other uses (Gregersen et al. 1994: 139).

3. This includes work within the Forest Margins Benchmark of southern Cameroon, household demographic surveys in the Center and South Provinces, a household farming systems survey in the East Province, and remote sensing work covering parts of the Center and East Provinces.

4. The Forest Margins Benchmark, a large area encompassing 15,400 km^2, has been defined over a gradient of population pressure in southern Cameroon that ranges from 5 to over 100 persons/km^2. The variation in forest condition and agricultural intensification permit the testing of a large range of hypothesized rural-forest interactions.

5. The irreversibility hypothesis of tropical deforestation is being tested by IITA in southern Cameroon, where various approaches for overcoming the agronomic and socioeconomic constraints impeding the conversion of deforested short fallow lands into multi-product, multi-strata cocoa agroforests (a combination of cocoa, exotic and indigenous fruit trees, and timber species) are being implemented in farmers' fields.

6. The current monitoring of the world's forests provided by FAO is too coarse to provide accurate measures of change. It is based on national level assessments of often dubious quality for fixed points in time that are then adjusted using a deforestation model, developed to permit the correlation of forest cover change over time with ancillary variables, including population change and density, initial forest cover and the ecological zone of the forest area under consideration (FAO 1997: 199).

7. Not the timber sector as one might expect on viewing the ubiquitous logging trucks cluttering Cameroon's road system.

8. MINEF (1995) cites a value of 50 billion CFA francs (US$83 million).

9. Minimum exploitable diameter (MED) has been established as the main tool for regulating timber fellings, with serious legal repercussions for breaking the rules (up to US$20,000 and 3 years in prison). A study conducted by CIRAD-Foret (1997) in the East Province found that about half of the trees that meet the MED standard are not of sufficiently good quality to be cut.

10. This, as Eba'a Atyi (1998) points out, explains the inconsistency between the 57 percent of the area designated for forest exploitation by nationals and the relatively low share of logs produced by them (36 percent).

11. In 1988–89 it was estimated that the extent of this illegal timber felling was 1 million m³, roughly half the legally felled timber (*Afrique Agriculture* 1990); MINEF (1995) estimates the volume at 150,000 cubic meters. Most of this timber is felled from lands held under customary tenure rights. Legally this timber belongs to the state, but farmers will often sell standing timber to local chainsaw operators who fell the tree and saw it into planks in situ. Because farmers have no legal right to these trees, they readily accept only a fraction of the value of the timber at international prices. The retail prices of these tropical hardwoods in Yaoundé are similarly much below what one would expect to pay. The outcomes of these informal markets are driven by poorly defined property rights, which should be addressed by appropriate policy interventions to legitimize the trade and encourage more efficient processing.

12. Eba'a Atyi (1998) reports an installed roundwood capacity of 1,988,000 m³ for 1996, assuming a complete log ban were implement a doubling of processing capacity to 4,000,000 m³ would be required.

Chapter 2

13. This policy sought to shift lending emphasis from industrial plantations under public sector control to activities related to smallholders and rural communities (including fuelwood), institution building, and environmental protection. Twice, the Bank missed the opportunity to use this project to apply the 1978 policy prescription: first at the original design in 1981, and second in the redesigned version of 1985. In both cases, the Bank stuck with the old approach of industrial plantations in the public sector.

14. Republic of Cameroon: Third Structural Adjustment Credit, Release of the Second Tranche (IDA/SecM99-396. June 24, 1999)

15. In principle, a competitive auction also limits the extent to which companies can exploit strategically the informational advantage they have on the government. They would have to bear the transaction costs associated with the assessment of the economic value of the concession under bidding.

16. One key assumption underlying the design of auction systems is homogeneity of the potential buyers with respect to the ability to pay. Such a bidding system is supposed to exploit differences in willingness to pay among bidders. Milgrom (1989:9) notes that there is no guarantee that the equilibrium outcome associated with the first-bid auction will be efficient. He further explains that in any environment where bidders have different observable characteristics, sealed bidding would produce an inefficient outcome with positive probability.

17. Communication with Country Team.

18. The Country Team explains that, so far, there is no evidence of collusion. While we do not dispute that statement, we think that policymakers should be aware of this potential problem. Particularly in an environment where the distinction between *national* and *foreigner* can be quite fuzzy. As stated earlier in connection with the structure of the logging industry in Cameroon, when faced with the difficulties of raising capital or operating in the export market, many Cameroonian entrants have no choice but to sell or lease their logging rights to a well-capitalized, foreign-owned concessionaire. Yet, one key assumption in optimal auction theory relates to the seller's ability to prevent resale among bidders after the auction. When this assumption fails, the optimality of the revenue-maximizing procedure may be lost (See Ausubel and Cramton 1998).

19. This mechanism is known to be compatible with truthful revelation of willingness to pay. This suggestion by Karsenty (1998) is further strengthened by the fact that the Vickrey auction is not distorted by resale, at least in a private-value setting (Ausubel and Cramton 1998).

20. See Dixit (1996:85–102) for details. Note also that this principle underlies tranching in the context of a structural program.

21. In principle, stumpage values can be estimated from the market price of logs net of transportation and logging costs. In practice, this is difficult, as such values would vary widely with species, grade, and distance from market. (Grut, Gray, and Egli 1991).

22. There are distortions associated with domestic processing requirements. The more valuable, high-quality species or grades are exported, while less valuable, low-quality ones (smaller and defective species, grades, or logs) end up being processed locally. This explains in part why processing is inefficient

and so wasteful. Such distortions are not the basis upon which to build a forest industry. Value-based log export taxes provide a more neutral, less distortionary domestic processing incentive because they raise the log export cost of all logs proportionally (Grut, Gray, and Egli 1991:31). They generate government revenues rather than encouraging inefficient behavior in producing sawn wood or plywood wastefully, or rent-seeking activity in acquiring export quotas.

23. These authors further explain that an export tax is more flexible than an export ban. It can be set to achieve any level of processing incentive. The level of incentive can be adjusted over time in response to tariffs imposed by the industrialized countries on processed products from developing countries. It can also be adjusted as the domestic industry develops and matures and incentives are less needed.

24. *Ordonnace No. 99/001 du 31 août 1999 complétant certaines dispositions de la loi no94/01 du 20 janvier 1994 portant régime des forêts, de la faune et de la pêche.*

25. At higher log prices, more species, logs, and trees would be used. Furthermore, the use of export taxes as instruments of domestic processing policy get more complex as one moves from logs to processed products. The domestic processing impacts of an export tax on logs are relatively clear and predictable, those on an export tax on products are not (Grut, Gray, and Egli 1991:30).

26. Fertilizers are only used on coffee in Cameroon.

27. Imports by coffee cooperatives in 1992 were only 1/6th (12,000 t vs. 63,000) their 1988 level (Abbot and Lloyd 1992).

28. In Cameroon, a 50 percent reduction in yield due to cocoa blackpod disease is commonly cited, which is above the 10–20 percent yield reduction experienced by Ghana and Côte d'Ivoire producers. This is in part due to the more virulent Phytophthora megakarya fungal agent, which is rampant in Cameroon and is threatening the industry in Côte d'Ivoire and Ghana, where the less virulent P. palmivora still predominates.

29. Based on annual production of 50,000 tons.

30. Women farmers account for more than half of the agricultural labor force in the humid forests of Cameroon and are the major producers of food crops (with the exception of plantain). An IITA study of the Forest Margins Benchmark institutions found that despite women's prominent role in agriculture, only 3 percent of extension agents in the benchmark were women (IITA unpublished data).

31. Before the liberalization of input markets, the Ministry of Agriculture had performed this task gratis.

32. The work is based on minutes of meetings and debates at the National Assembly, documents from the MINEF, and interviews with Members of Parliament, civil servants at the MINEF and the President's office, World Bank staff at the Resident Mission, and a national logger.

33. Ekoko (1998) cites observers who saw a conflict of interests in this arrangement, as future business opportunities for this firm depended on the content of the new law.

34. This is not the case in countries such as Brazil, Costa Rica, Indonesia, and China covered by this OED review.

35. See Dixit (1996:9).

Bibliography

Abbot, R., and D. Lloyd. 1992. "Privitization of Fertilizer Marketing in Cameroon: A Fourth-Year Assessment of the Fertilizer Sub-Sector Reform Program." Report to the AMIS project, USAID/Cameroon.

Afrique Agriculture. 1990. "Des mesures radicales pour redresser les filières dans un contexte de contraintes." *Afrique Agriculture* 175: 30–38.

Afrique Agriculture. Various years. Paris.

Afrique Agriculture Forêt. 1990. "La forêt camerounaise sous haute convoitise." *Afrique Agriculture Forêt* 175: 90–106.

Armstrong, Robert P. 1996. *Ghana Country Assistance Review: A Study in Development Effectiveness.* Operations Evaluation Department. Washington, D.C.: World Bank.

Angelsen, A., and D. Kaimowitz. 1998. *When Does Technological Change Promote Deforestation? Theoretical Approaches and Some Empirical Evidence.* Bogor: Center for International Forestry Research.

ASB (Alternatives to Slash and Burn). 2000. *Phase II Final Report for the Forest Margins Benchmark of Cameroon.* Nairobi: ASB Coordination Office.

Asheim, Geir B. 1994. *Sustainability: Ethical Foundations and Economic Properties.* World Bank Policy Research Working Paper 1302. Washington, D.C.

Ausubel, Lawrence M., and P. Cramton. 1998. *The Optimality of Being Efficient: Designing Auctions.* World Bank Policy Research Working Paper 1985. Washington, D.C.

Bayalama, Sylvain. 1995. "Deforestation and Development in the Congo Basin." Ph.D dissertation, University of Denver, Denver, Colorado. Photocopy.

Bawag Besong, Joseph. 1992. "New Directions in National Forestry Policies: Cameroon." In Kevin Cleaver, M. Munasinghe, M. Dyson, N. Egli, A. Peuker, and F. Wencélius, eds., *Conservation of West and Central African Forests.* Washington, D.C.: World Bank.

Bawag Besong, Joseph, and F. L. Wencélius. 1992. "Realistic Strategies for Conservation of Biodiversity in the Tropical Moist Forests of Africa: Regional Overview." In Kevin Cleaver, M. Munasinghe, M. Dyson, N. Egli, A. Peuker, and F. Wencélius, eds., *Conservation of West and Central African West Forests.* Washington, D.C.: World Bank.

Binswanger, Hans, and K. Deininger. 1997. *Explaining Agricultural and Agrarian Policies in Developing Countries.* World Bank Policy Research Working Paper 1765. Washington, D.C.

Binswanger, H., and P. Pingali. 1987. "Technology Priorities for Farming in Sub-Saharan Africa." *World Bank Research Observer* 3.

Binswanger, H. P., and V. Ruttan. 1978. *Induced Innovation: Technology, Institutions and Development.* Baltimore: The Johns Hopkins University Press.

Bowles, Ian A., et al. 1998. "Logging and Tropical Forest Conservation." *Science* 280: 1899–1900.

Bromley, Daniel W., and M.M. Cernea. 1989. *The Management of Common Property Natural Resources: Some Conceptual and Operational Fallacies.* World Bank Discussion Paper 57. Washington, D.C.

Brunner, Jake, and Francois Ekoko. 2000. "Cameroon Case Study." In Frances Seymour and Navroz Dubash, eds., *The Right Conditions: The World Bank, Structural Adjustment, and Forest Policy Reform.* Washington, D.C.: World Resources Institute.

Bulow, Jeremy, and J. Roberts. 1989. "The Simple Economics of Optimal Auctions." *Journal of Political Economy* 97(5): 1060–90.

Campbell, Donald E. 1995. *Incentives: Motivation and Economics of Information. Cambridge*, U.K.: Cambridge University Press.

Carret, Jean-Christophe. 1998. *La réforme de la fiscalité forestière au Cameroon: Contexte, bilan et questions ouvertes.* Paris: Centre d'économie industrielle, Ecole Nationale Supérieure des Mines de Paris.

CIRAD-Forêt. 1997. "Le projet d'Aménagement Pilote Intégré de Dimako." Yaoundé: Ministère de l'Environnemnt et de Forêts.

Clayton, Anthony M.H., and N. J. Radcliffe. 1996. *Sustainability: A Systems Approach.* London: Earthscan.

Cleaver, K. 1993. *A Strategy To Develop Agriculture in Sub-Saharan Africa and a Focus for the World Bank.* World Bank Technical Paper 203, Africa Technical Department Series. Washington, D.C.

Cleaver, Kevin, and G. Schreiber. 1992. "Population, Agriculture, and the Environment in Africa." *Finance and Development* (June): 34–35.

Coase, Ronald Harry. 1990. *The Firm, the Market and the Law.* Chicago: The University of Chicago Press.

Collier, Paul. 1997. "Redesigning Conditionality." *World Development* (U.K.)25:1399–407.

Collier Paul, Patrick Guillaumont, Sylviane Guillaumont, and Jan Willem Gunning. 1997. "Redesigning Conditionality." *World Development* 25 (9):1399–407.

Corden, Max W. 1987. *Protection and Liberalization: A Review of Analytical Issues.* International Monetary Fund Occasional Paper No. 54 (August). Washington, D.C.

Côté, S. 1993. "Plan zonage du Cameroun du Meridional." Agence Canadienne de Développement International et Ministère de l'Environnement et des Forêts.

Diaw, C. 1997. "Si, Nda bot and Ayong. Shifting Cultivation, Land Uses and Property Rights in Southern Cameroon." IITA Humid Forest Station, Yaoundé. Photocopy.

Diaw, Marieuw Chimere. 1997. *Si, Nda Bot and Ayong: Shifting Cultivation, Land Use and Property Rights in Southern Cameroon.* London: Rural Development Forestry Network.

Dixit, Avinash K. 1996. *The Making of Economic Policy: A Transaction-Cost Politics Perspective.* Cambridge, MA: MIT Press.

Dixit, Avinash K., and Barry Nalebuff. 1991. *Thinking Strategically: The Competitive Edge in Business, Politics, and Everyday Life.* New York: Norton.

Dixon, R. K., J. A. Perry, E. L. Vanderklein, and F. Hiol. 1996. "Vulnerability of Forest Resources to Global Climate Change: Case Study of Cameroon and Ghana." *Climate Research* 6: 127–33.

Dongmo, J. L. 1981. *Le dynamisme Bamiléké* (Cameroun). Volume I: *La Maîtrise de l'Espace Agraire.* Yaoundé: Université de Yaoundé, Centre d'Edition et de Production pour l'Enseignement et la Recherche.

D'Silva, Emmanuel H., and D. Kariyawasam, eds. 1995. *Emerging Issues in Forest Management for Sustainable Development in South Asia: Proceedings of a South Asia Seminar.* Manila: Asian Development Bank.

Dury, S. 1997. *Les comportements d'épargne des ménages ruraux: Spécification d'un modèle dynamique et estimation sur données d'enquêtes (Java-Indonésie).* Montpellier, France: INRA Série Etudes et Recherches n° 108.

Dyson, Mary. 1992. "Concern for Africa's Forest Peoples: A Touchstone of a Sustainable Development Policy." In Kevin Cleaver, M. Munasinghe, M. Dyson, N. Egli, A. Peuker, and F. Wencélius, eds., *Conservation of West and Central African Forests.* Washington, D.C.: World Bank.

Eba'a Atyi, R. 1998. *Cameroon's Logging Industry: Structure, Economic Importance and Effects of Devaluation.* Center for International Forestry Research Occasional Paper Number 14. Center for International Forestry Research, Indonesia.

Ekoko, Francois. 1998. *Environmental Adjustment in Cameroon: Challenges and Opportunities for Policy Reform in the Forest Sector.* Washington, D.C.: WRI.

———. 1997. "The Political Economy of the 1994 Cameroon Forestry Law." Paper presented at the African Regional Hearing of the World Commission on Forests and Sustainable Development in Yaoundé, May 1997. Center for International Forestry Research, Cameroon.

FAO (Food and Agricultural Organization). 1997. *State of the World's Forests.* Oxford, UK: Words and Publications.

———. 1988. ''Plan d'Action Forestier Tropical.'' Volume I : *Résumé exécutif, Rapport de mission conjointe interagence de planification et de revue pour le secteur forestier.* Rome: PNUD-FAO.

Findlay, R. 1991. "The New Political Economy: Its Explanatory Power for LDCs." In G. Meier, ed., *Politics and Policy-Making in Developing Countries.* San Francisco: ICS.

Findlay, R., and Stanislaw Wellisz. 1983. "Some Aspects of the Political Economy of Trade Restrictions." *Kyklos* 36(3): 469–81.

Frischtak, Leila L. 1994. *Governance Capacity and Economic Reform in Developing Countries*. World Bank Technical Paper 254. Washington, D.C.

Garbus, L., A. Pritchard, and O. Knudsen. 1991. *Agricultural Issues in the 1990s: Proceedings of the Eleventh Agricultural Sector Symposium*. Washington, D.C.: World Bank.

Gartlan, Stephen. 1992. "Practical Constraints on Sustainable Logging in Cameroon." In Kevin Cleaver, M. Munasinghe, M. Dyson, N. Egli, A. Peuker, and F. Wencélius, eds., *Conservation of West and Central African Forests*. Washington, D.C.: World Bank.

Gartlan, S. 1989. "La Conservation des Ecosystèmes forestiers du Cameroun." UICN, Alliance mondiale pour la nature. Commission des Communautés Européennes.

Gaston, G., S. Brown, M. Lorenzini, and K.D. Singh. 1998. "State and Change in Carbon Pools in the Forests of Tropical Africa." *Global Change Biology* 4(1): 97–114.

Global Environment Facility (GEF). 1998a. *Summary Report: Study of GEF Project Lessons*. Washington, D.C.

———. 1998b. *GEF Lessons Notes* (July). Washington, D.C.

———. 1997. *Project Implementation Review of the Global Environment Facility*. Washington, D.C.

———. 1996a. *Operational Strategy*. Washington, D.C.

———. 1996b. *The GEF Project Cycle*. Washington, D.C.

———. 1996c. *Incremental Costs*. Washington, D.C.

———. 1996d. *A Framework of GEF Activities Concerning Land Degradation*. Washington, D.C.

Gockowski, James, and Doyle Baker. 1996. "An Ecoregional Methodology for Targeting Resource and Crop Management Research in the Humid Forest of Central and West Africa." Paper presented at 1996 Biennial Meeting of Rockefeller Social Science Research Fellows, 15–17 August, 1996, Nairobi, Kenya.

Gockowski, J., and S. Dury. 1999. "The Economics of Cocoa-Fruit Agroforests in Southern Cameroon." Paper presented to the International Workshop on Multi-Strata Systems with Perennial Tree Crops, CATIE, Costa Rica, 22–25 February (forthcoming in *Agroforestry Systems*).

Gockowski J., and M. Ndoumbé. 1999. *The Economic Analysis of Horticultural Production and Marketing in the Forest Margins Benchmark of Southern Cameroon.* IITA RCMR Monograph No. 27. Ibadan Nigeria.

Gockowski, J., B. Nkamleu, and J. Wendt. 1999. "Implications of Resource Use Intensification for the Environment and Sustainable Technology Systems in the Central African Rainforest." Invited paper, American Agricultural Economics Association International Pre-Conference on Agricultural Intensification, the Environment, and Rural Development, July 30 to August 1, 1998, Salt Lake City, Utah.

Gockowski, J., D. Baker, J. Tonye, S. Weise, M. Ndoumbé, T. Tiki-Manga, and A. Fouaguégué. 1998. "Characterization and Diagnosis of Farming Systems in the ASB Forest Margins Benchmark of Southern Cameroon." Yaoundé, IITA Humid Forest Ecoregional Center. Photocopy.

Greg, G., B. Sandara, L. Massimiliano, and K. D. Singh. 1998. *State and Change in Carbon Pools in the Forest of Tropical Africa. Global Change Biology.* Viale delle Terme di Caracalla, Italy: U.S. EPA National Health and Environmental Effects Research Laboratory, Western Ecology Division, and FAO Forestry Department.

Gregersen, Hans, J. Spears, and B. Belcher. 1994. "Containing Deforestation: The Policy Challenge." *Quarterly Journal of International Agriculture* 33(2): 138–49.

Grut Mikael, John Gray, and Nicolas Egli. 1991. *Forest Pricing and Concession Policies: Managing the High Forests of West and Central Africa.* World Bank Technical Paper 143. Washington, D.C.

Hartwick, John M. 1993. "Forestry Economics, Deforestation, and National Accounting." In Ernst Lutz, ed., *Toward Improved Accounting for the Environment.* Washington, D.C.: World Bank.

Hazell, Peter, and W. Magrath. 1992. "Summary of World Bank Forestry Policy." In Kevin Cleaver, M. Munasinghe, M. Dyson, N. Egli, A. Peuker, and F. Wencélius, eds., *Conservation of West and Central African Forests.* Washington, D.C.: World Bank.

Hodge, Ian. 1995. *Environmental Economics.* New York: St. Martin's.

Horta, Korinna. 1997. "La Nuit Coloniale continue à porter son Ombre Immense sur ce Vaste Continent: Cameroon's War Against Subsistence. A Socio-Economic and Ecological Analysis Against the Background of the Chad–Cameroon Oil-Pipeline." *The BOS NiEuWSLETTER* 16(3:37): 66–75.

IIED (Institut International pour l'Environnement et le Développement). 1987. "Le territoire Forestier Camerounais: les ressources, les intervenants, les politiques d'utilisation."

IUCN (World Conservation Union). 1994. *Forest Atlas of Africa.* Gland, Switzerland.

Jayaraja, Carl, et al. 1996. *Social Dimensions of Adjustment: The World Bank Experience, 1980–1993.* Washington, D.C.: World Bank.

Jordan, Lisa. 1997. *Sustainable Rhetoric vs. Sustainable Development: The Retreat from Sustainability in World Bank Development Policy.* Washington, D.C.: World Bank.

Kaimowitz, D., and A. Angelsen. 1997. *A Guide to Economic Models of Tropical Deforestation.* Jakarta: CIFOR.

Karsenty, Alain. 1998. *Environmental Taxation and Economic Instruments for Forestry Management in the Congo Basin.* London: IIED.

Karsenty, A., and D.V. Joiris. 1999. *Les systèmes locaux de gestion dans le basssin Congolais.* Libreville, Gabon: Central African Regional Program for the Environment (CARPE), CARPE-IR1.

Kishor, Nalin, and L. Constantino. 1994. "Sustainable Forestry: Can It Compete." *Finance and Development* 31:36–39.

Klein, Martha, and Mark van der Wal. 1997. "About Tropical Hardwood, Chocolates and Gorillas: Conservation of Forest Fauna in South Cameroon." *The BOS NiEuWSLETTER* 16(3:37): 50–58.

Kotto-Same, J., P.L. Woomer, A. Moukam, and L. Zapfack. 1997. "Carbon Dynamics in Slash-and-Burn Agriculture and Land Use Alternatives of the Humid Forest Zone in Cameroon." *Agriculture, Ecosystems & Environment* 65(3):245–56.

Leffler, Keith B., and R.R. Rucker. 1991. "Transaction Costs and the Efficient Organization of Production: A Study of Timber-Harvesting Contracts." *The Journal of Political Economy* 99(5): 1060–87.

Lemmens, R.H., and M.S. Sosef. 1997. "The Flora of the Congo Basin." *The BOS NiEuWSLETTER* 16(3:37): 21–25.

Letouzey, Rene. 1986. *Manual of Forest Botany: Tropical Africa.* Nogent-sur-Marne, France: Centre technique forestier tropical.

Manyong, V. M., J. Smith, G.K. Weber, S.S. Jagtap, and B. Oyewole. 1996a. "Macro-Characterization of Agricultural Systems in Central Africa: An Overview." IITA Resource and Crop Management Research Monograph No. 22. Ibadan, Nigeria.

Manyong, V.M., J. Smith, G.K. Weber, S.S. Jagtap, and B. Oyewole. 1996b "Macro-Characterization of Agricultural Systems in West Africa: An Overview." IITA Resource and Crop Management Research Monograph No. 21. Ibadan, Nigeria.

Mba'zoa, and J. Gockowski. 1999. *Nutrition and Consumption Habits of the Urban Poor in Central Africa: A Components Analysis of Sauce in Yaoundé Cameroun.* International Institute of Tropical Agriculture.

Mertons, B., and Eric F. Lambim. 1997. "Spatial Modeling of Deforestation in Southern Cameroun: Spatial Disaggregation of Diverse Deforestation Processes." *Applied Geography* 17 (2).

Milgrom, Paul. 1989. "Auctions and Bidding: A Primer." *Journal of Economic Perspectives* 3(3):3–22.

Millington, A.C., and K. Pye. 1994. *Environmental Change in Drylands: Biogeographical and Geomorphological Perspectives.* Chichester, U.K., New York: J. Wiley & Sons.

Millington, A.C., R.W. Critchley, T.D. Douglas, and P. Ryan. 1994. *Estimating Woody Biomass in Sub-Saharan Africa.* Washington, D.C.: World Bank.

MINAGRI (Ministere de l'agriculture). 1994. *Evaluation de l'impact du PNVFA* (Rapport provisiore). Yaoundé, Cameroon, Direction de l'agriculture.

———. 1988. *Résultats de l'enquête agricole campagne 1986-1987.* Yaoundé, Cameroon, Ministere de l'agriculture, Direction des enquêtes agro-economiques et de la planification agricole.

———. 1987. *Agricultural census—traditional sector.* Volumes 2.E and 2.F (Results of the Center and South Provinces). Yaoundé.

MINEF (Ministère de l'environnement et des forêts). 1996. *Plan National de gestion de l'environnement au Cameroun (PNGE):* Volume II: *Analyses sectorielles.* Yaoundé.

———. 1995. *Plan d'action forestière.* Vol. II: *Rapport de Synthèse.* Yaoundé.

Monke, E. A., and S.R. Pearson. 1989. *The Policy Analysis Matrix for Agricultural Development.* Ithaca and London: Cornell University Press.

Munasinghe, Mohan, ed. 1996. *Environmental Impact of Macroeconomic and Sectoral Policies.* Washington, D.C.: World Bank.

Nashashibi, K., and S. Bazzoni. 1994. "Exchange Rates Strategies and Fiscal Performance in Sub-Saharan Africa." *IMF Staff Papers* 41(1):76–122.

Ndjatsana, M., and Nga T. Ndjodo. 1998. "L'exploitation de l'Azobe au Cameroun: Etude de cas de la Société Design." Séminaire camerouno-hollandais CIPRE/ICCO sur l'exploitation et la gestion durable de l'Azobe.

Ndoye, O. 1995a. "Commercialization and Diversification Opportunities for Farmers in the Humid Forest Zone of Cameroon: The Case of Non-Timber Forest Products." Consultancy report for ASB Project Cameroon. Yaoundé, IITA-Humid Forest Station.

————. 1995b. "The Markets for Non-Timber Products in the Humid Forest Zone of Cameroon and Its Borders: Structure, Conduct, Performance and Policy Implications." Yaoundé, Cameroon, CIFOR/IITA.

Newman, Kate. 1992. "Forest People and People in the Forest: Investing in Local Community Development." In Kevin Cleaver, M. Munasinghe, M. Dyson, N. Egli, A. Peuker, and F. Wencélius, eds., *Conservation of West and Central African Forests*. Washington, D.C.: World Bank.

Newman, Kate, and W. Cruz. 1995. *Economy-wide Policies and Environment: Lessons from Experience*. World Bank Environment Paper 10. Washington, D.C.

Ntangsi, J. 1987. *The Political and Economic Dimensions of Agricultural Policy in Cameroon*. World Bank MADIA Paper 673-04. Washington, D.C.

O'Halloran, Eavan, and Vicente Ferrer. 1997. "The Evolution of Cameroon's New Forestry Legal, Regulatory, and Taxation System." World Bank, Washington, D.C. Photocopy.

Okigbo, B. N. 1994. "Conservation and Use of Germoplasm in African Traditional Agriculture and Land Use System." In A. Putter, ed., *Safeguarding the Genetic Basis of Africa's Traditional Crops*. CTA, the Netherlands/IPGRI, Rome.

Peters, Charles M. 1996. *The Ecology and Management of Non-Timber Forest Resources*. Washington, D.C.: World Bank.

Pezzey, John. 1992. *Sustainable Development Concepts: An Economic Analysis*. World Bank Environment Paper 2. Washington, D.C.

Picciotto, R. 1992. *Participatory Development: Myths and Dilemmas*. World Bank Working Paper 930. Washington, D.C.

Picciotto, R., and E. Wiesner. 1998. *Evaluation and Development: The Institutional Dimension*. Washington D.C.: World Bank.

Plouvier, Dominiek. 1997. "Short Overview of the Situation of Tropical Moist Forests and Forest Management in Central Africa and Markets for African Timber." *The BOS NiEuWSLETTER* 16(3:37): 42–49.

Plouvier, Dominiek, and J.L. Roux. 1997. "Promotion of Sustainable Forest Management and Certification in Timber Producing Countries of West and Central Africa." *The BOS NiEuWSLETTER* 16(3:37): 99–108.

Porter, Gareth, et al. 1998. *A Study of GEF's Overall Performance.* Washington, D.C.: Global Environment Facility.

Prince, S. D, and S. N. Goward. 1995. "Global Primary Production: A Remote Sensing Approach." *Journal of Biogeography* 22: 815–35.

Reardon, T., and S. A. Vosti. 1995. "Links Between Rural Poverty and the Environment in Developing Countries: Asset Categories and Investment Poverty." *World Development* 23(9): 1495–506.

Redwood, John III, R. Robelus, and T. Vetleseter. 1998. *Natural Resource Management Portfolio Review.* World Bank Environment Department Paper 58. Washington, D.C.

Ross, Michael. 1996. "Conditionality and Logging in the Tropics." In Robert O. Keohane and M.A. Levy, eds., *Institutions for Environmental Aid: Pitfalls and Promise.* Cambridge, MA: MIT Press.

Ruf, F. 1995. *Booms et crises du cacao, les vertiges de l'or brun.* Paris: Editions Kharthala.

Salanié, Bernard. 1998. *The Economics of Contracts: A Primer.* Cambridge, MA: MIT Press.

Sandler, Todd. 1997. *Global Challenges: An Approach to Environmental, Political, and Economic Problems.* Cambridge, U.K.: Cambridge University Press.

Satabié, V. 1995. "The Biodiversity of Cameroonian Flora." Paper presented at the Workshop on Biodiversity Prospecting for Cameroon, Madagascar, and Ghana, April 24-May 2, Cameroon, by Bioresources Development and Conservation Programme, USAID Washington.

Sayer, Jeffrey. 1992. "Development Assistance Strategies to Conserve Africa's Rainforests." In Kevin Cleaver, M. Munasinghe, M. Dyson, N. Egli, A. Peuker, and F. Wencélius, eds., *Conservation of West and Central African Forests.* Washington, D.C.: World Bank.

Serageldin, I. 1991. *Saving Africa's Rainforests.* Washington, D.C.: World Bank.

Serageldin, I., and A. Mahfouz, eds. 1995. *The Self and the Other: Sustainability and Self-Empowerment.* Washington, D.C.: World Bank.

Sharma, Narendra P., S. Rietbergen, C.R. Heimo, and J. Patel. 1994. *A Strategy for the Forest Sector in Sub-Saharan Africa*. World Bank Technical Paper 251. Washington, D.C.

Shepherd, G., D. Brown, M. Richards, and K. Schreckenberg. 1998. *The EU Tropical Forestry Sourcebook*. London: Overseas Development Institute.

Shirley, Mary M., and L. Colin Xu. 1997. *Information, Incentives, and Commitment: An Empirical Analysis of Contracts between Governments and State Enterprises*. World Bank Policy Research Working Paper 1769. Washington, D.C.

Smith, J., and G. Weber. 1991. "Strategic Research in Heterogenous Mandate Areas: An Example from the West African Savanna." In J. Anderson, ed., *Agricultural Technology Issues for the International Community*. Washington, D.C.: CAB/World Bank.

Stiglitz, J.E. 1996. "The Role of Government in Economic Development." In Michael Bruno and B. Pleskovic, eds., *Annual World Bank Conference on Development Economics*. Washington, D.C.: World Bank.

———. 1989. "Economic Organization, Information, and Development." In Hollis Chenery and T.N. Srinivasan, eds., *Handbook of Development Economics*, Vol. I. Amsterdam: Elsevier.

Struhsaker, T.T. 1998. "A Biologist's Perspective on the Role of Sustainable Harvest in Conservation." *Conservation Biology* 12(4): 930–32.

Sunderlin, W. D., and J. Pokam. 1998. "Economic Crisis and Forest Cover Change in Cameroon: The Roles of Migration, Crop Diversification, and Gender Division of Labor." CIFOR. Photocopy.

Sunderlin, William D., J. Pokam, and K. Wadja. 1998. *L'Impact de la Crise Économique sur les Populations, les Migrations et el Couvert forestier du Sud-Cameroun*. CIFOR Occasional Paper 25.

TAC/CGIAR. 1993. *The Ecoregional Research Approach to Research in the CGIAR*. Rome: TAC Secretariat.

Thenkabail, P. S. 1999. "Characterization of the Alternative to Slash-and-Burn Benchmark Research Area Representing the Congolese Rainforests of Africa Using Near-Real-Time SPOT HRV Data. *International Journal of Remote Sensing* 20 (5): 839–77.

Varlet, F., and D. Berry. 1997. *Réhabilitation de la protection phytosanitaire des cacaoyers et caféiers du Cameroun*. Tome 1: *Rapport principal*, Tome 2: *Annexes*. CIRAD et CICC, Rapport No. 96/97/SAR.

Verdoes, Arie. 1997. "Congo Basin Regional Profile." *The BOS NiEuWSLETTER* 16(3:37): 5–12.

Vivien, J., and J. Faure. 1985. *Arbres des Forêts denses d'Afrique Centrale*. Paris. Agence de Coopération Culturelle et Technique.

Vosti, S. A., and others. 1998. *Intensifying Small-Scale Agriculture in the Western Brazilian Amazon: Issues, Implications and Implementation*. Washington, D.C.: International Food Policy Research Institute.

von Amsberg, Joachim. 1998. "Economic Parameters of Deforestation." *World Bank Economic Review* 12(1): 133–53.

Vooren, A.P. 1992. "Harvest Criteria for Tropical Forest Trees." In Kevin Cleaver, M. Munasinghe, M. Dyson, N. Egli, A. Peuker, and F. Wencélius, eds., *Conservation of West and Central African Forests*. Washington, D.C.: World Bank.

Warford, J.J., M. Munasinghe, and W. Cruz. 1997. *The Greening of Economic Policy Reform*, Volume I: *Principles*. Washington, D.C.: World Bank.

Winterbottom, Robert. 1992. "Tropical Forestry Action Plans and Indigenous People: The Case of Cameroon." In Kevin Cleaver, M. Munasinghe, M. Dyson, N. Egli, A. Peuker, F. Wencélius, eds., Conservation of West and Central African Forests. Washington, D.C.: The World Bank.

World Bank. 1999. *Republic of Cameroon: Country Status Report*. Prepared for the June 1999 SPA Meeting. Washington, D.C.

———. 1998/99. *World Development Report: Knowledge for Development*. New York: Oxford University Press.

———. 1998a. *Cameroon: Country Assistance Strategy—Progress Report*. Washington, D.C.

———. 1998b. *Assessing Development Effectiveness: Evaluation in the World Bank and the International Finance Corporation*. Operations Evaluation Department. Washington, D.C.

———. 1998c. *1997 Annual Review of Development Effectiveness*. Operations Evaluation Department. Washington D.C.

———. 1997. *Five Years after Rio: Innovations in Environmental Policy*. Environmentally Sustainable Development Studies and Monographs Series. Washington, D.C.

———. 1996a. *Resettlement and Development: The Bank-Wide Review of Projects Involving Involuntary Resettlement, 1986–1993*. Environment Department. Washington, D.C.

————. 1996b. *Maximizing Sustainable Forestry Development Impact in the Countries of the Congo Basin: Strategy for AF3.* AF3AE. Washington, D.C.

————. 1995a. *Mainstreaming Biodiversity in Development: A World Bank Assistance Strategy for Implementing the Convention on Biological Diversity.* Environment Department Paper 029. Washington, D.C.

————. 1995b. *Republic of Cameroon: Biodiversity Conservation and Management.* GEF. Washington, D.C.

————. 1995. *Working with NGOs: A Practical Guide to Operational Collaboration between the World Bank and Nongovernmental Organizations.* Operations Policy Department. Washington, D.C.

————. 1994a. *Conditional Lending Experience in World Bank–Financed Forestry Projects.* Operations Evaluation Department Report 13820. Washington, D.C.

————. 1994b. *Global Environmental Facility: Independent Evaluation of the Pilot Phase.* Washington, D.C.

————. 1993. *Poverty Reduction Handbook.* Washington, D.C.

————. 1992. *Guidelines for Monitoring and Evaluation of GEF Biodiversity Projects.* Environment Department. Washington, D.C.

————. 1991a. *The Forest Sector: A World Bank Policy Paper.* Washington, D.C.

————. 1991b. *Forestry: The World Bank's Experience.* Operations Evaluation Department. Washington, D.C.

————. 1990. *World Development Report 1990: Poverty.* Washington, D.C.

————. 1989. *Renewable Resource Management in Agriculture.* Operations Evaluation Department Report No PUB7345. Washington, D.C.

WRI (World Resources Institute). 1994. *World Resource 1994-95.* New York: Oxford University Press.

Young, H. Peyton. 1994. *Equity: In Theory and Practice.* Princeton, NJ: Princeton University Press.

Operations Evaluation Department Publications

The Operations Evaluation Department (OED), an independent evaluation unit reporting to the World Bank's Executive Directors, rates the development impact and performance of all the Bank's completed lending operations. Results and recommendations are reported to the Executive Directors and fed back into the design and implementation of new policies and projects. In addition to the individual operations and country assistance programs, OED evaluates the Bank's policies and processes.

Summaries of studies and the full text of the Précis and Lessons & Practices can be read on the Internet at http://www.worldbank.org/html/oed.

How To Order OED Publications

Operations evaluation studies, World Bank discussion papers, and all other documents are available from the World Bank InfoShop.

Documents listed with a stock number and price code may be obtained through the World Bank's mail order service or from its InfoShop in downtown Washington, D.C. For information on all other documents, contact the World Bank InfoShop.

For more information about this study or OED's other evaluation work, please contact Elizabeth Campbell-Pagé or the OED Help Desk.

Operations Evaluation Department
Partnerships & Knowledge Programs (OEDPK)
E-mail: ecampbellpage@worldbank.org
E-mail: eline@worldbank.org
Telephone: (202) 473-4497
Fax: (202) 522-3200

Ordering World Bank Publications

Customers in the United States and in territories not served by any of the Bank's publication distributors may send publication orders to:

The World Bank
P.O. Box 960
Herndon, VA 20172-0960
Fax: (703) 661-1501
Telephone: (703) 661-1580

The address for the World Bank publication database on the Internet is: http://www.worldbank.org (select publications/project info).

E-mail: pic@worldbank.org
Fax: (202) 522-1500
Telephone: (202) 458-5454

The World Bank InfoShop serves walk-in customers only. The InfoShop is located at:

701 18th Street, NW
Washington, DC 20433, USA

All other customers must place their orders through their local distributors.

Ordering by E-mail

If you have an established account with the World Bank, you may transmit your order by electronic mail on the Internet to: books@worldbank.org. Please include your account number; billing and shipping addresses; and the title, order number, quantity, and unit price for each item.